Calamity Theory

Forerunners: Ideas First

Short books of thought-in-process scholarship, where intense analysis, questioning, and speculation take the lead

FROM THE UNIVERSITY OF MINNESOTA PRESS

Joshua Schuster and Derek Woods
Calamity Theory: Three Critiques of Existential Risk

Daniel Bertrand Monk and Andrew Herscher
The Global Shelter Imaginary: IKEA Humanitarianism and Rightless Relief

Catherine Liu
Virtue Hoarders: The Case against the Professional Managerial Class

Christopher Schaberg
Grounded: Perpetual Flight . . . and Then the Pandemic

Marquis Bey
The Problem of the Negro as a Problem for Gender

Cristina Beltrán
Cruelty as Citizenship: How Migrant Suffering Sustains White Democracy

Hil Malatino
Trans Care

Sarah Juliet Lauro
Kill the Overseer! The Gamification of Slave Resistance

Alexis L. Boylan, Anna Mae Duane, Michael Gill, and Barbara Gurr
Furious Feminisms: Alternate Routes on *Mad Max: Fury Road*

Ian G. R. Shaw and Marv Waterstone
Wageless Life: A Manifesto for a Future beyond Capitalism

Claudia Milian
LatinX

Aaron Jaffe
Spoiler Alert: A Critical Guide

Don Ihde
Medical Technics

Jonathan Beecher Field
Town Hall Meetings and the Death of Deliberation

Jennifer Gabrys
How to Do Things with Sensors

Naa Oyo A. Kwate
Burgers in Blackface: Anti-Black Restaurants Then and Now

Arne De Boever
Against Aesthetic Exceptionalism

Steve Mentz
Break Up the Anthropocene

John Protevi
Edges of the State

Matthew J. Wolf-Meyer
**Theory for the World to Come: Speculative Fiction
and Apocalyptic Anthropology**

Nicholas Tampio
Learning versus the Common Core

Kathryn Yusoff
A Billion Black Anthropocenes or None

Kenneth J. Saltman
The Swindle of Innovative Educational Finance

Ginger Nolan
The Neocolonialism of the Global Village

Joanna Zylinska
The End of Man: A Feminist Counterapocalypse

Robert Rosenberger
Callous Objects: Designs against the Homeless

William E. Connolly
**Aspirational Fascism: The Struggle for Multifaceted
Democracy under Trumpism**

Chuck Rybak
UW Struggle: When a State Attacks Its University

Clare Birchall
**Shareveillance: The Dangers of Openly Sharing and
Covertly Collecting Data**

la paperson
A Third University Is Possible

Kelly Oliver
Carceral Humanitarianism: Logics of Refugee Detention

P. David Marshall
The Celebrity Persona Pandemic

(Continued on page 127)

Calamity Theory
Three Critiques of Existential Risk

Joshua Schuster and
Derek Woods

University of Minnesota Press

MINNEAPOLIS

LONDON

ISBN 978-1-5179-1291-8 (PB)
ISBN 978-1-4529-6658-8 (Ebook)
ISBN 978-1-4529-6700-4 (Manifold)

Published by the University of Minnesota Press, 2021
111 Third Avenue South, Suite 290
Minneapolis, MN 55401-2520
http://www.upress.umn.edu

Available as a Manifold edition at manifold.umn.edu

The University of Minnesota is an equal-opportunity educator and employer.

Contents

Introduction: What Is Existential Risk?   1

1. Endgame Philosophy   23

2. Probability and Speculation   45

3. The Existential Roots of Existential Risk   79

Conclusion: Opening the "Letter from Utopia"   115

Acknowledgments   125

Introduction: What Is Existential Risk?

DO WE LIVE IN EXISTENTIALLY CRITICAL TIMES? Based on the recent upsurge in the use of the word from Right and Left political spectrums, existential sentiment is common and frequently exaggerated.[1] The term "existential" has become a household word again, but has its philosophical meaning changed or transformed with its renewed popularization? In a sweeping judgment on existential risks in the present, Stephen Hawking pronounced in 2016 that "this is the most dangerous time for our planet."[2] Hawking points to the rise of several different technologies that could either unleash doomsday levels of violence or bring about a new kind of human flourishing by transforming the structures of human existence.

1. For example: Donald Trump's use of language "poses an existential threat to democracy," in George Lakoff and Gil Duran, "Trump Has Turned Words into Weapons," *Guardian,* June 13, 2018, https://www.theguardian.com/commentisfree/2018/jun/13/how-to-report-trump-media-manipulation-language; Julia Hollingsworth, "Climate Change Could Pose 'Existential Threat' by 2050: Report," *CNN,* June 4, 2019, https://www.cnn.com/2019/06/04/health/climate-change-existential-threat-report-intl/index.html.
2. Stephen Hawking, "This Is the Most Dangerous Time for Our Planet," *Guardian,* December 1, 2016, https://www.theguardian.com/commentisfree/2016/dec/01/stephen-hawking-dangerous-time-planet-inequality.

Existential thought has always foregrounded as a "first philosophy" an urgent examination of who or what kind of existences have their lives at risk. But is it an exaggeration to say that the entirety of human existence itself is at stake today? Is this a time when whole realms of existentiality are at stake, such that forms of existence—and not only human existence—are at risk of being irrevocably changed or destroyed? If extinction is unquestionably on the horizon, is there no alternative to the pervasive apocalyptic mode?

We open with these perilous questions in order to understand how they have become foundational in the quickly growing academic field of existential risk, the study of species-wide extinction threats to human existence, including ecological calamities, nuclear weapons, viral pandemics, disastrous biotechnologies and nanotechnologies, extraterrestrial threats from asteroids to aliens, and superintelligent computers either malevolent to humans or benevolent but rendering humans displaceable and disposable. These extinction scenarios are not all new to the twenty-first century, but the planetary scale, speed, and power of human activities that could exceed human control and precipitate an extinction event are converging at an accelerated rate.[3] Because there are so many globally transformative technologies and social behaviors putting into question the future of life on the planet, humans collectively are reaching a turning point in which the decisions we make in "the next 50 years will determine the next 10,000 years."[4]

3. Extinction as a philosophical problem is not new either. See John Leslie, *The End of the World: The Science and Ethics of Extinction* (New York: Routledge, 1996).

4. Mats Anderson et. al., "Global Challenges Annual Report Ambassador's Preface," in *Global Catastrophic Risks 2017,* ed. Nick Bostrom and Milan Ćirković (Stockholm: Global Challenges Foundation, 2017), 10. This sense of temporality as intertwined with a "tipping point" in the coming decades is also central to the environmental thought and action needed to grapple with the impacts of global warming and ecological distress caused in the Anthropocene. Will Steffen, Paul Crutzen, and John R. MacNeil, "The Anthropocene: Are Humans Now Overwhelming the Great Forces of Nature?," *Ambio* 36, no. 8 (December 2007): 620.

But what methods should we use to comprehend existential risks? How has this field of study been shaped, and by whom? What does existential thinking mean today, with the generation of postwar existentialist philosophers receding in the rearview mirror? What kinds of existentiality or ways of being are emerging or becoming submerged in the wake of these intensified risks? How are some ways of life more existentially at risk than others, and how should one respond to the unevenly distributed harms that are immediately endangering some existences? How are we to think the existential risks and ruins of recent making—the genocides, extinctions, and exhaustions of our planet—yet still find a way to think and act that is not wholly consumed by apocalyptic thought? As Donna Haraway asks, "How can we think in times of urgency *without* the self-indulgent and self-fulfilling myths of apocalypse, when every fiber of our being is interlaced, even complicit, in the webs of processes that must somehow be engaged and repatterned?"[5]

We offer here a critical reflection on the rise in existential risk thinking as a new form of calamity theory, in order to understand its upsurge in popularity, its foundational arguments and assumptions, and the negative implications of this kind of risk analysis. Primarily, we analyze the initial arguments for existential risk posited by the analytic philosopher Nick Bostrom, whose essays from the past two decades have played the most prominent role in launching this movement. While we refer to other existential risk analysts throughout, we focus on Bostrom's work because he is the founder of the field and remains the most widely cited and provocative thinker of existential risk. He can well serve as a synecdoche of the field, but our focus also expands to thinkers like Toby Ord, Martin Rees, and Phil Torres, and to the field's considerable impact in popular media.[6] Existential risk has featured in TED Talks viewed in

5. Donna Haraway, *Staying with the Trouble: Making Kin in the Chthulucene* (Durham: Duke University Press, 2016), 35.

6. Other notable work in the field of existential risk includes Max Tegmark, *Life 3.0: Being Human in the Age of Artificial Intelligence* (New York: Vintage, 2017); Sam Harris, *Making Sense: Conversations on*

the millions and in Bostrom's *Superintelligence: Paths, Dangers, Strategies* (2014), a *New York Times* bestseller, as well as significant media coverage from *Wired*, the *Guardian*, *The Economist*, and the *New Yorker*. *Three Critiques* is the first book to provide a close examination of how the field became constituted primarily along Bostrom's terms, and also the first to develop a sustained critique using recent methodologies from the environmental humanities and science and technology studies.

Bostrom defines an existential risk as "one where an adverse outcome would either annihilate Earth-originating intelligent life or permanently and drastically curtail its potential."[7] We will examine closely this particular definition of existential risk, which indicates that human extinction would only be one kind of existential imperilment and includes other kinds of loss of "potential" as just as disastrous as extinction. Most dramatically, this definition does not actually name human extinction as the primary catastrophe (note humans are not specifically mentioned) but rather "Earth-originating intelligent life" as the fundamental unit of existential value that must not be curtailed or destroyed.

Bostrom's definition of existential risk has had an outsized impact across the field. His notion of existential risk is already popular in many sectors of philosophy, especially in AI safety research and the effective altruism movement. The new field has accrued mil-

---

Consciousness, Morality, and the Future of Humanity (New York: Ecco, 2020); Seth D. Baum and Bruce E. Tonn, eds., "Special Issue on Confronting Future Catastrophic Threats to Humanity," *Futures* 72 (September 2015): 1–96. Thomas Moynihan provides an insightful intellectual history of existential risk in *X-Risk: How Humanity Discovered Its Own Extinction* (Falmouth: Urbanomic, 2020).

7. Nick Bostrom, "Existential Risks: Analyzing Human Extinction Scenarios and Related Hazards," *Journal of Evolution and Technology* 9, no. 1 (2002): 2. Note that all essays by Bostrom cited here are taken from his website nickbostrom.com. Bostrom has helpfully offered open access to his work. For many of these essays, the pdfs supplied do not correspond in page numbers to the original publications. We cite the page numbers from the essays as presented from the website because they are widely available to the public.

lions of views on social media platforms and podcasts, but a wider discussion of how existential risk thinking has spread in these public and academic forums has yet to occur. We are conscious of addressing the unusual crossover of readers versed in the environmental humanities (including extinction studies) and readers more comfortable with existential risk as an offshoot of analytic philosophy, probabilistic risk theory, and transhumanist advocacy. Because existential risk is almost unstudied in the environmental humanities, we do not take for granted that all critics will have the same problems with it that we do.

Existential Risk as a Philosophical Field

The phrase "existential risk" dates in its rise of popularity to Bostrom's 2002 essay, "Existential Risks: Analyzing Human Extinction Scenarios and Related Hazards." Bostrom's essay, written in a clear, crisp, and brisk style, cuts through some of the bewildering and oracular rhetoric of apocalyptic thinking to make human extinction a topic for analytic and utilitarian philosophy. Using dispassionate prose, Bostrom eschews long-winded diatribes and fire-and-brimstone moralizing, presenting a brief survey of possible species-level threats. He offers a minimalist guideline for ethics in the face of such risks—thin on specifics and any self-critique. In a further remarkable move, Bostrom aligns the study of existential risk with a vision of the complete overcoming of existential problems, pointing to the apparent resolution of such risks in the potential "astronomical" value of massive technological transformations to come that would permit humans or Earth-originating intelligence to spread across the galaxy.[8]

When he wrote "Existential Risks," Bostrom had been known primarily as an enthusiastic proponent of transhumanism.[9] With the

8. Nick Bostrom, "Astronomical Waste: The Opportunity Cost of Delayed Technological Development," *Utilitas* 15, no. 3 (2003): 308–14.
9. Nick Bostrom, "Transhumanism—An Idea Whose Time Has Come," in *Doctor Tandy's First Guide to Life Extension and Transhumanity* (Palo

popularizing of the phrase "existential risk," which he defines in his initial essay as "global, terminal risks" to humanity, Bostrom soon garnered widespread acclaim, leading to his professorship at Oxford in 2003 and his directorship of the Future of Humanity Institute in 2005.[10] In 2012, Cambridge University established the Centre for the Study of Existential Risk, and a number of other institutions have followed, with significant private donor backing from technocratic elites such as Elon Musk.[11] The Future of Humanity Institute received 13.3 million pounds in funding from the Open Philanthropy Project in 2018, an effective altruism group primarily funded by Dustin Moskovitz, who cofounded Facebook with Mark Zuckerberg and three others. Skype cofounder Jaan Tallinn is also cofounder of the Boston-area Future of Life Institute, a think tank with ties to MIT devoted to studying AI safety and other technological extinction risks. Musk sits on the FLI scientific advisory board, and on a similar board at the Cambridge Centre. Support for existential risk studies from deep pockets has become part of the story of the field. For us, what is important is how such extraordinary funding affects the methods and vision of the philosophers in the field, who consistently focus their attention on wondrous technological feats as the antidote to existential problems and generally eschew any concrete social or environmental justice initiatives, including those

Alto: Ria University Press, 2001). For critical studies of the social history and conceptual basis of transhumanism, see Abou Farman, *On Not Dying: Secular Immortality in the Age of Technoscience* (Minneapolis: University of Minnesota Press, 2020); Andrew Pilsch, *Transhumanism: Evolutionary Futurism and the Human Technologies of Utopia* (Minneapolis: University of Minnesota Press, 2017).

10. Bostrom, "Existential Risks," 1. For The Future of Humanity Institute, see https://www.fhi.ox.ac.uk/. The institute's website hosts a helpful collection of publications in the field of existential risk.

11. For the Centre for the Study of Existential Risk, see https://www .cser.ac.uk/. For a survey of institutional trends, see Seán Ó Héigeartaigh, "The State of Research in Existential Risk," *First International Colloquium on Catastrophic and Existential Risk,* ed. B. John Garrick (B. John Garrick Institute for the Risk Sciences, 2017), 37–52.

that contest the tech corporations hungry for the resources and energies that fuel the artificial intelligence boom.

As it stands, we might say that there are three categories of research on existential risk: (1) The self-conscious members of the field, a comparatively small but well-funded group of academics trained in analytic philosophy and risk modeling (along with a handful of scientists such as Rees and Max Tegmark). This is where the field has more depth than breadth. (2) The wider, popular media circuits in which the specialist claims of existential risk analysis circulate, which include a broad range of influential magazines and news sites. Outside its academic publications, the field has wide and enviable impact, but not so much depth. (3) The work of researchers, mostly scientists and data analysts, who publish things that thinkers like Bostrom consider relevant to existential risk. Here there is undefinable breadth and depth, but because of how easy it is to make the case for relevance to a topic as expansive as existential risk, we cannot cover this third category in detail. To clarify, our critique is directed at the first two categories both because we think the field and its impact are problematic and because these are theoretically instructive case studies for environmental humanists interested in the methods and impacts of conceptualizing extinction.

Some of the exigency of our project is spurred by the widespread popularity of Bostrom's launching of the field under his terms, but we are also wary about how existential risk theory has been conducive to a warm reception by a "tech bro" Silicon Valley audience. Existential risk discourse currently thrives amid high capitalist fever for AI and transhumanist life extension start-ups. Slavoj Žižek acutely articulates the Silicon Valley ideology whereby triumphing over the "impossible" becomes the paramount focus of every company, while improving just a little bit the social welfare of the nation, let alone the planet, is deemed truly inconceivable: "You want to raise taxes a little bit for the rich; they tell you it's impossible [because we] lose competitivity. You want more money for health care, they tell you 'impossible; this means a totalitarian state.'

There's something wrong in the world, where you are promised to be immortal but cannot spend a little bit more for healthcare."[12] Instead of the hard work of building a safer world together through universalizing norms of respect, dignity, and consent across varied ways of existential flourishing and social solidarity, Bostrom offers a list of calamitous scenarios and transhuman rescue fantasies that aim to eventually pan out in "masters of the universe" powers. As it is currently constructed, the field of existential risk ultimately reinforces anthropocentric thinking, cheers the capitalist acceleration of AI (as long as it builds in moral safeguards), elevates individualistic egoic satisfactions as the truth of "astronomical" values and the destiny of technology, and flees at all costs from the existential commons of cross-species vulnerability and finitude.

The institutional history of existential risk thinking is quite banal in comparison to the dystopian feverishness and utopian bravado implicit in these arguments. But the references and style of argument in this field play a crucial role in determining what counts as an existential risk, who is most immediately exposed to life threats, and what "existential" means. It's already a remarkable phenomenon to observe a new cross-disciplinary field being founded in its initial stages, but the stakes of this particular field are legitimately hair-raising and epochal. This field has not only turned some academics into major public figures, it has won them a seat at the table as consultants on some of the most important decisions being made by companies and governments that will affect everyone.

The researchers contributing to the field of existential risk, Bostrom included, have stressed adamantly the need for more scholarship and funding to study extinction scenarios and probabilities of civilizational collapse. Yet this field has cultivated a noticeable lack of diversity of references, voices, and participants so far, with existential risk scholars being mostly white males from wealthy

12. Slavoj Žižek, "Don't Fall in Love with Yourselves," in *Occupy! Scenes from an Occupied America,* ed. Astra Taylor, Keith Cessen, and editors from N+1 (London: Verso, 2011), 69.

Western countries and well-funded foundations and universities. These countries, institutions, and identities are not facing immediate existential risks. While we do not deny that all humans everywhere might face sudden extinction threats at any time, what stands out is how little attention existential risk theorists are paying to marginalized peoples and colonized existences that face the end of their way of life on a daily basis. Also troubling is the lack of engagement with the long history of research and testimony of those whose lives and cultures have been threatened with destruction. The vast majority of scholarship in the field of existential risks ignores reams of scholarship and personal accounts of living through mass suffering and extreme violence, preferring instead to jump to speculative and spectacular scenarios summed up in bite-sized terms and axioms.

Environmental Humanities and Risk Criticism

In parallel to the field of existential risk, new approaches to the study of extinction are growing in the humanities, emphasizing environmental justice and multispecies storytelling.[13] The history of

13. Deborah Bird Rose, *Wild Dog Dreaming: Love and Extinction* (Charlottesville: University of Virginia Press, 2011); Claire Colebrook, *Death of the PostHuman: Essays on Extinction,* vol. 1 (Open Humanities Press, 2015); Ursula Heise, *Imagining Extinction: The Cultural Meanings of Endangered Species* (Chicago: University of Chicago Press, 2016); Thom van Dooren, *Flight Ways: Life and Loss at the Edge of Extinction* (New York: Columbia University Press, 2014); Deborah Bird Rose, Thom van Dooren, and Matthew Chrulew, eds., *Extinction Studies: Stories of Time, Death, and Generations* (New York: Columbia University Press, 2017); Audra Mitchell, "Beyond Biodiversity and Species: Problematizing Extinction," *Theory, Culture and Society* 33, no. 5 (2016): 23–24; Ashley Dawson, *Extinction: A Radical History* (New York: OR Books, 2016); Richard Grusin, ed., *After Extinction* (Minneapolis: University of Minnesota Press, 2018); Susan McHugh, *Love in a Time of Slaughters: Human-Animal Stories against Genocide and Extinction* (University Park: Penn State University Press, 2019). See also the special issue "Writing Extinction," *Oxford Literary Review* 41, no. 1 (2019), ed. Sarah Wood.

understanding extinction is tied to humanistic inquiry about how to live together and share a world and what happens when worlds in common are lost. Extinction is a biological fact and also conceptual crisis that can potentially put all of our concepts and values into question, because these too are at stake in the disappearance of ways of life. Extinction entails the existential end of a life form, while also compelling us to think about the existentiality of the terms we use to understand and respond to extinction. Examining the existentiality of concepts and terms involves considering how the methods we use run into their own limits and crises in thinking extinction. In the eloquent phrasing of Matthew Chrulew and Rick De Vos,

> Extinction challenges our thinking and writing. Such overwhelming disappearance of ways of being, experiencing and making meaning in the world disrupts familiar categories and demands new modes of response. It requires that we trace multiple forms of both countable and intangible loss, the unravelling of social and ecological communities as a result of colonialism and capture, development and defaunation and other destructive processes. It brings forth new modes of commemoration and mourning, and new practices of archiving and survival. It calls for action in the absence of hope.[14]

Thinking extinction also entails registering the impact of extinction on how we think. One must find a way to address the end of address for whichever life form, including the human, might disappear. As we show, this incorporation of the finitude of thought into philosophy is a hallmark of existential thought. This paradox, using thought to imagine the end of thought, has produced a tradition of philosophical inquiry generative of new ways to imagine existential risks.

Understanding any extinction event requires attention to the specificities of embodiment, care, communal memory, power over life and death, collective mourning, evolutionary theory, and ecological connectedness that allow us to cognize both world building and world loss. Yet, as we will see, concepts like these that have

14. Matthew Chrulew and Rick De Vos, "Extinction: Stories of Unraveling and Reworlding," *Cultural Studies Review* 25, no. 1 (2019): 23.

become central to comprehending extinction in the environmental humanities have garnered little attention so far from the field of existential risk. Advocates of existential risk philosophy are preoccupied foremost with tallying a range of immediate and far-term doomsday scenarios. Thinkers in this field employ a version of philosophical rationalism, statistical analysis, and speculation on the future of risks to human survival. But these are not the only ways of understanding extinction risks, and not the only practices and discourses of existential urgency.

In what ways are existential risk and environmental humanities complimentary, and in what ways do they markedly diverge? How might both fields contribute to what Molly Wallace calls a "risk criticism"? Risk criticism combines both cultural and technological reflections on risk toward developing not just urgently needed foresight, but also a sense of "precautionary reading" that "emphasizes its uncertainness, unpredictability, and incalculability."[15] Precautionary reading links to Wendy Hui Kyong Chun's argument that climate change disaster scenarios use "hypo-real" prediction—that is, a modeling between hypothesis and reality—to help avoid the outcome they predict. "If models work properly as evidence, they become unverifiable: if we are convinced of their verisimilitude, we will act in such a way that their predicted results can never be corroborated by experience."[16] In existential risk analysis, the primary aim is to come up with a projection in which uncertainty becomes a

15. Molly Wallace, *Risk Criticism: Precautionary Reading in an Age of Environmental Uncertainty* (Ann Arbor: University of Michigan Press, 2016), 18–20. The term "risk" in existential risk is largely taken from risk analysis methods in economics, social health, and game theory. The use of the term in the humanities owes some debt to Ulrich Beck, *Risk Society: Towards a New Modernity,* trans. Mark Ritter (London: Sage, 1992).

16. Wendy Hui Kyong Chun, "On Hypo-Real Models or Global Climate Change: A Challenge for the Humanities," *Critical Inquiry* 41, no. 3 (Spring 2015): 678. Frédéric Neyrat argues that capitalism in the time of existential risk has created "clairvoyance societies" that "take the future as their point of application: to predict the future in order to avoid it." *Atopias: Manifesto for a Radical Existentialism* (New York: Fordham University Press, 2017), 15.

practical tool used to foreclose the disastrous futures it anticipates. "Precautionary" existential risk analysis requires the humanist tools of associative, critical, and creative speculative thinking as well as data gathering and rationalist rulemaking.

Existential Re-Mix

In yet another strange omission, none of the philosophers who are shaping the field reflect on the conceptual history of the term "existential." They are existential thinkers without thinking existentialism. None seem to care that existential philosophy reached its intellectual peak precisely as a response to the existential risks of the first half of the twentieth century, forcing philosophy to reckon with scientific evidence of species extinction, world wars, genocides, concentration camps, nuclear weapons, imperialism, and structurally oppressive systems of racism and sexism. Existential thought still provides relevant and useful tools to reflect on existential risks. The methodological shortcomings of existential risk should not be excused in the name of epochal urgency. As we will argue, these preferences and limitations go to the heart of the stakes, reasons, and practical outcomes of thinking existential crises today.

We are not subscribers to any school of existentialism. Our claim is rather that the contributions of existential thought to the field of philosophy—specifically the emphasis on a need for both subjective and objective accounts of subjectivity and the becoming of knowledge, following Hegel's phrasing that "everything turns on grasping and expressing the True, not only as *Substance,* but equally as *Subject*"[17]—provide crucial insights for comprehending existential risks. However, we do not define human existence on terms typically associated with existential-humanism. For us, the human is a "cybernetic triangle"[18] of animal-human-machine,

17. G. W. F. Hegel, *Phenomenology of Spirit,* trans. A. V. Miller (Oxford: Oxford University Press, 1977), 10.
18. Dominic Pettman, *Human Error: Species-Being and Media Machines* (Minneapolis: University of Minnesota Press, 2011), 5.

terms that both conjoin and mutually oppose each other and yet together compose the phenomena of human life intertwined with other kinds of existence.[19] We offer an initial definition of (unaffiliated) existential thought based on three core claims: (1) Existential thought thinks the structural conditions and phenomenal "lived experience" of everyday existence; (2) existential thought thinks the limit-experiences of existence, that is, the extremes of existence, including mortality and extinction, as well as phase transitions from one form of existence to another; and (3) existential thought implicates the thinker in the thought, such that the participation of the thinker has an effect on the thought. We find that existential risk overwhelmingly borrows from the second definition while neglecting the first and the third. Yet the analysis of "lived experience" and embodiment as expansive for thought (not something subdued by probabilistic reasoning or superseded by enhanced intelligence) is central to what the term "existential" means and so also to our understanding of existential limit risks. Uncoupling one definition of "existential" from the others will have profound consequences, as we demonstrate later.

Bostrom makes it clear that he views the restriction or demise of *intelligent life,* not necessarily human life as such, as the utmost calamity. Bostrom prioritizes expanding "the space of reasons" toward "astronomical" values as the broadest horizon for his philosophy of existential risk. This view of expansive reasoning, superintelligence, and galactic-scale value premised on maximizing these apparent utilitarian goods shapes significantly how existential risks are defined and assessed. In Bostrom's thinking, because existential risks are endemic to humanity and have no permanent solutions, we have a duty to not only expand our philosophical knowledge and rational

19. Many use the term "posthuman" for this animal–human–machine complex. See Rosi Braidotti, *The Posthuman* (New York: Polity, 2013). We do not use the term posthuman in the way Braidotti does (though we largely agree with her work); for the sake of being consistent with Bostrom, we adopt his use of posthumanity as synonymous with transhumanity.

discourse on existential risk but also pursue modifications to our existential condition. These modifications can range from enhancing intellectual, biological, and technological powers to improve our ability to mitigate existential risks and to maximize the utilitarian value of any number of future lives.

Existential thought is concerned with the philosophical basis for expanding the possibility and accessibility of world-making and world-sharing across the space of existences. In the face of rising existential risks and contestations over definitions of "existential" and "risk," we reaffirm the existential condition for *expanding ways of flourishing together and sharing existence on Earth,* which aligns with the aims of environmental justice as "the search for fair ways of sharing environmental burdens and benefits and collectively creating a future in which the dignity and rights of all people are respected."[20] Expanding the space of reasons is not the only criterion for truth.[21] Expanding the space of existence toward ends of environmental justice includes creating broader conditions of existential flourishing for diverse ways of life on a shared planet. The space of existence includes the range of embodied affects, experiences, material and ecological relations—these all contribute to expanding the space of thought and existentiality, which is not reducible to intelligence or utilitarian calculations of present and future value. To paraphrase Kant, reasons without existence are empty, while existence without reasons is blind.

What does this argument for expanding the space of existence have to do with existential risk? To summarize our reply:

20. Elizabeth Ammons and Modhumita Roy, "Introduction," *Sharing the Earth: An International Environmental Justice Reader* (Athens: University of Georgia Press, 2015), 2.

21. David Roden argues that different iterations of humanism, posthumanism, and transhumanism correspond to different physical and metaphysical "possibility spaces" (53). The human possibility space is not the same as the posthuman possibility space, and while there would be much overlap, some posthumans would experience and know things that humans could never understand, and presumably vice versa. David Roden, *Posthuman Life: Philosophy at the Edge of the Human* (New York: Routledge, 2015).

1. All the existential risks facing humans and other life forms on Earth concern what Hannah Arendt called the fundamental necessity to "share the Earth."[22] In our view the existential condition is the ecological condition. Extinction and ecology are intertwined phenomena; extinction makes no sense without a context in which the singularity of a life form makes a unique and irreplaceable contribution to expanding the space of existences. There is no such thing as ecology without processes of natality and mortality, speciation and extinction, situated in specific habitats. Too much extinction destroys ecology, but no extinction at all is another way to eliminate the ecological condition. The concepts, methodology, and mitigation of existential risks need to build in this intertwined existential and ecological condition rather than seek to dismantle this connection in the name of some speculative immunity.

2. All decisions regarding the permanent change of the existential condition for humans or nonhuman animals should involve consultation and consent of all species existences. This is clearly an ideal scenario that is not currently feasible, but it should be built into the methodology of existential risk studies. We have no current political models for this form of radical politics. We have no form of address in which to address all existences. Yet any change to the human existential condition would involve all current and future humans, whether or not they consent, and would impact the vast majority of animal life as well. We will need to imagine and institute new forms of address, consent, respect, and inclusion—a process that none of the existing literature in the field of existential risk prioritizes.

3. Bostrom's theorizations and remedies for existential risk divide up the present and future of humanity along the restrictive lines of a hierarchical valuation of intelligence. Bostrom claims that existential risks show us that enhanced intelligence toward a

---

22. Hannah Arendt, *Eichmann in Jerusalem: A Report on the Banality of Evil* (New York: Penguin, 2006), 279. Arendt's reflections are influenced by Kant's statement in "Perpetual Peace" that everyone has the "right to the earth's surface which the human race shares in common." Immanuel Kant, *Political Writings,* 2nd ed., trans. H. B. Nisbet (Cambridge: Cambridge University Press, 1991), 106.

transhuman condition is not just a possibility and a favorable option, but a moral and ontological obligation. Bostrom consistently conflates the obligation to prevent human extinction with the obligation to develop humanity toward transhumanism. This conflation is predicated on an evaluative hierarchy of superintelligence that places other forms of embodied and ecological intelligence into subordinate positions (this is Claire Colebrook's critique, discussed in chapter 3). In Bostrom's view, humans can modify the animal–human–machine existential condition toward one that abandons the animal part and accelerates the human–machine complex toward a more promising future. Our claim is that the animal–human–machine condition functions as a package deal, such that expanding the space of intelligence is connected to expanding the space of existential flourishing. We doubt that it is either possible or desirable to peel existence away from the evolutionary trajectories and ecological interdependencies of life.

4. The existential condition is the basis of our elementary commonalities with all other life on Earth. Existential risk analysis should not be about abandoning, sterilizing, dominating, or solving once and for all the existential commons but finding ways to flourish together. Such existential commons are the basis for our intersubjective sociality with human and nonhuman life, what N. Katherine Hayles, writing on the novel coronavirus in 2020, has called "*humans as species in common*" and "*humans in biosymbiosis.*"[23] Any change to this structure would reverberate across the planetary biosymbiotic commons, permanently changing those commons and perhaps revoking the conditions of possibility for having a shared planet at all. Our position aligns with what Deborah Bird Rose advocates as "ecological existentialism"[24] and Stacy Alaimo's call for an ethics and politics based on grappling with the exposure and entanglement of our bodies with other bodies. Alaimo writes, "Performing exposure as an ethical and political act means to reckon with—rather than disavow—such

23. N. Katherine Hayles, "Novel Corona: Posthuman Virus," *Critical Inquiry* (blog), April 17, 2020, https://critinq.wordpress.com/2020/04/17/novel-corona-posthuman-virus/.

24. Rose, *Wild Dog Dreaming*, 44.

horrific events and to grapple with the particular entanglements of vulnerability and complicity that radiate from disasters. . . . To occupy exposure as insurgent vulnerability is to perform material rather than abstract alliances, and to inhabit a fraught sense of political agency that emerges from the perceived loss of boundaries and sovereignty."[25]

Bostrom envisions a scenario in which existential risks eventually give way to the posthuman prosperity of enhanced cognitive capabilities. But the current existential structure we have already allows for different forms of first-person experiences, self-transformations, and collective flourishing. We are not rejecting Bostrom's posthumanism out of a humanist fear of transformation or nostalgia for a simpler human condition, but with a sense of solidarity and collective possibility toward the future, which can include new technologies but need not rely on such technologies to save us. As Cary Wolfe and Zakiyyah Iman Jackson have argued, posthumanism can be a mutation of ideas and practices that disrupts existing racialized and speciesist norms of being human, not some effort to transcend our finite bodies and ecologies.[26] We do not need any special new technology or enhancement in order to cultivate Rose's "ecological existentialism" and Alaimo's "insurgent vulnerability."

The Three Critiques

Each of this book's three chapters is a distinct critique of existential risk: a discursive analysis of the field's foundational terms, a review of the use of science and probability theory from the stance

25. Stacy Alaimo, *Exposed: Environmental Politics and Pleasures in Posthuman Times* (Minneapolis: University of Minnesota Press, 2016), 5.

26. Cary Wolfe, *What Is Posthumanism?* (Minneapolis: University of Minnesota Press, 2009); Zakiyyah Iman Jackson, *Becoming Human: Matter and Meaning in an Anti-Black World* (New York: New York University Press, 2020).

of critical science and technology studies, and an examination of the existential implications of existential risk.

Chapter 1 looks at the language, methodology, and rationale of Bostrom's first major essay on existential risk, published in 2002. This essay remains foundational for the field, especially in its terminology and broad overview of what constitutes an existential risk. We highlight in particular how Bostrom's philosophical style makes heavy use of typologies and lists. Typologies are not forms conducive to philosophical debate and self-critical argumentation; they are brief summarizing statements designed to be read with speed. These typologies betray an impatience with traditional philosophizing, scholarship, and interpretive reading across the humanities and social sciences. We go into further detail examining how Bostrom constructs his list of existential risks, and how he reckons that existing institutional safeguards against such risks should be bolstered yet not really trusted. His conclusion that no current form of security would likely be adequate opens the door to authoritarian preventive actions and fantasies of total behavioral control, a position that bodes poorly for this philosophy and speaks to the problems at the roots of this field.

Chapter 2 provides a reflection on the use of risk analysis and probability thinking in the development of existential risk theory. This chapter draws on science and technology studies to critique the use of science, and the tendency toward scientism, in existential risk theory, while raising questions about the field's relation to science fiction. Existential risk analysts seek to estimate the chance that a given cause will lead to the extinction of humanity or even life in general, using probability models that are generally Bayesian. However, these theorists also recognize that since there has never been a recorded existential risk event (extinction event) for humans, probability runs up against an impossible data point. The crux of existential risk analysis is the synthesis of these two elements. But this crux is also a failure: as we show in this section, existential risk analysis is *constituted* by a radical mismatch between method and object of study. This chapter also examines more closely the claim that existential risk work is scientific and not science fiction, which ignores the theoretical and predictive insights available in the imaginative work of literary speculation.

Chapter 3 provides a more expanded history and definition of existential thought. We claim that many philosophers associated with existentialism, and many subsequent philosophers critical of these initial existential positions, provide a compelling reflection on why existential thought and existential risk must be coupled together. We also discuss why not one single philosopher in the field of existential risk has put any effort into reflecting on the history and insights of what "existential" means. This chapter examines more closely three rejections of Bostrom-style existential risk philosophizing, first in the work of contemporary theorists Claire Colebrook and Ray Brassier, and then in the historical contributions of Hans Jonas. The chapter closes with a consideration of what kind of scale critique is most effective for existential ecological thought, turning to questions of cosmology and eschatology that are often cryptically at stake in studies of existential risk. Finally, the book's conclusion examines the rhetorical lures of Bostrom's utopian writings, specifically his "Letter from Utopia."

We base our arguments in an ecological–existential framework, but this is not a claim that we must forever leave nature alone; Bostrom is right to be skeptical that we must defer to "the natural order" as the sole standard of the good.[27] We do not view ecology and the existential condition as unchanging origins and fixed ends. Ecological existence is dynamic, made and mediated by biological relations, cultural practices, technological changes, and imaginative pursuits. We are already cyborgs, but what kinds and with what futures? We support technological improvements to alleviate suffering and improve the function of ecosystems. Yet we do not support the idea that all of the planet and all of the existential

27. Nick Bostrom, "In Defense of Posthuman Dignity," *Bioethics* 19, no. 3 (2005): 205.

condition should be under technological control and intelligent
enhancement in order to mitigate and master existential risks. We
are against species-wide irreversible technological changes, changes
that would be made without consultation or consent. We are against
changing the demos (the lived world) by nondemocratic means. The
planet already is a calamity for many species in recent centuries,
and many human communities exist on a daily basis in existentially
oppressive conditions. We want a better world through technology,
but not through technology alone, and not requiring a dependence
on the exploitation, extractivism, inequality, and massive economic
disparity that facilitates current hyperindustrial capitalism. And we
reject Bostrom's claim that transhuman enhancements will become
the new categorical imperative as long as they pose no existential
risks. While there is no metaphysical necessity to be drawn from
nature, there is also no metaphysical necessity for transhumanism.

These three critiques are not primarily moral criticisms. Rather
they are oriented around methodological and ontological disputes
with existential risk theory, although our points do have moral
consequences that we defend as implicated in our method. To put
our cards on the table, we envision the long-term tasks of humanity
as follows: share and tend the Earth, work toward symbiosis in the
sense of evolutionary mutualism and social collectivity, harness the
energy of the sun, and communicate with the galaxy. To advance
this vision, we will need new forms of address and belonging, as
well as new forms of accountability and collaborative exploratory
epistemologies, on a planetary scale. We will need solar energy
technologies and new solar cultures that cultivate solar commons
and collaborations rather than solar sovereigns and inequity ex-
acerbated by solar capitalism for the few. This vision informs our
methodological assessment of the long-term horizons of existential
risk and our own understanding of the existential condition.

This book, however, is unapologetically on the whole a work of
negative critique. Much of the work of theory remains unmasking,
demystification, and showing the unspoken historical and philosophical conditions of discourses such as existential risk—and

this work is inclusive of more affirmative and creative conceptualizations. If existential risk were only an obscure philosophical subfield, it might not warrant this treatment. Why not reason with "no-nonsense" straight talk about the mitigation of possible human extinction, using whatever methods seem right to intellectuals who want to go there? Since existential risk has risen to a fairly high level of popularity and media impact, the field deserves close scrutiny from critics with a background in the study of related topics across the sciences and humanities. We have not been the only ones, and we hope others will continue this effort in the future. We also acknowledge that negative critique has obvious limitations in terms of providing immediate relief or resolution for existential risks. While that is not our primary purpose here, we note that critique, as well as fostering existential commons and reasoning through practices of care, relation building, economic equity, and environmental attentiveness, model ways to share a world—thus mitigating existential risk. A combination of patience, critique, sociability, imagination, and scientific reason may take us further in crafting just and habitable worlds than wide-eyed apocalyptic doomscrolling or the fantasy of living "long enough to live forever," as Ray Kurzweil puts it.

Finally, there is a cosmological undercurrent in this book that overlaps with its interest in risk, apocalypticism, utopianism, and most of all, existential ecology. If there is one thing that mitigates our antagonism toward existential risk (beyond the obvious need to raise awareness about future horrors), it is this new excitement about cosmology. We don't know where it will go, but we have begun to wonder why there is so little interest in the topic in Left critical humanities circles. After all, one way of explaining existential risk's incongruous mixture of sci-fi fantasy plots and dry calculation is that it taps a deep vein of public desire for secular eschatology—that is, for stories and scenarios about the origins and ends of humanity, life, and the universe. Inquiries on ontological domains at vast scales and strange thresholds are not separate from attention to environmental and social justice on Earth, which include care for existential pluralism and respect for cross-cultural attunements to cosmological conjectures. Such a perspective requires cultivating what Marisol de la Cadena and Mario Blaser call a "pluriverse" predicated on "the practice of a world of many worlds, . . . heterogeneous worldings coming together as a political ecology of practices, negotiating their difficult being together in heterogeneity."[28] While cosmological concepts intertwined with existential risk are not this book's main strand, we have been glad to follow them in unexpected directions.

28. Marisol de la Cadena and Mario Blaser, "Introduction: Pluriverse, Proposals for a World of Many Worlds," in *A World of Many Worlds,* ed. Marisol de la Cadena and Mario Blaser (Durham, N.C.: Duke University Press), 4.

1. Endgame Philosophy

BOSTROM'S FOUNDATIONAL 2002 ESSAY on existential risk begins by setting the bar incredibly and unnecessarily high for a risk to be existential according to his "typology of risk."[1] The definitional standard for a danger to be existential is "one where an adverse outcome would either annihilate Earth-originating intelligent life or permanently and drastically curtail its potential" (ER, 2). Bostrom claims that a risk needs to be both global and terminal to constitute an existential risk. In this model, "global endurable risks" that would create mass suffering but not lead to human extinction do not constitute true existential risks. In Bostrom's typology, planetary-scale historically violent events are consigned to the less risky category of "catastrophe." A terminal existential event means either total extinction or some kind of irreversible change that would structurally prevent humans from achieving their collective potential. In this classification system, genocide is by definition not an existential risk since it is only supposedly localized at the level of the "genos" or a kind of human life, not the whole of humanity. According to Bostrom, "An example of a local terminal risk would be genocide leading to the annihilation of a people (this happened to several Indian *[sic]* nations)" (ER, 2). World wars, the enslavement of vast proportions of the world, and bacterial and viral epidemics like the black plague and AIDS (and COVID-19) also would not regis-

---

1. Bostrom, "Existential Risk," 1. Cited hereafter as ER.

ter in existential terms since neither the entirety of humanity nor the whole structure that facilitates human potential or "Earth-originating intelligence" are at risk. As Bostrom states, "Tragic as such events are to the people immediately affected, in the big picture of things—from the perspective of humankind as a whole—even the worst of these catastrophes are mere ripples on the surface of the great sea of life" (ER, 2).

Here we pause already on the first pages of Bostrom's essay to examine this controversial position. Perhaps the most startling aspect of Bostrom's initial essay on existential risk is not its wide-eyed openness to any species-wide crisis, real or speculative, but his insistence on categorizing the large-scale violent events of the past and present as not at all existential threats. By demarcating only extreme calamity to be truly existential in scale, Bostrom's claims for a strict existential threshold not only evinces obvious callousness, it sets the bar too high to be a useful measure. This model regretfully but purposefully ignores the violence and suffering inflicted on groups of people in which their existential condition is at stake. Bostrom assumes this violence has been and will continue to be contained as "endurable global risks since humanity could eventually recover" (2). This is dangerously close to rationalizing epochal histories of the suffering of minoritized and oppressed peoples for the sake of purported definitional consistency since something of "humanity" would survive. Furthermore, the threat of genocide and planet-sweeping violence still hangs over everyone and thus does bear directly on the existential condition of our species. While it may be the case that humanity as a whole is not imperiled immediately in the event of a particular genocide, that humans can intentionally commit such acts with explicit declarations that some lives are worth less than others indicates that there is nothing in the human condition that prevents humans from utterly destroying each other's humanity. Does it really need to be said that the "logic" of genocide qualifies as extinctionary, even if the results are often "incomplete"? Why has Bostrom then not built this recognition into his philosophy?

Genocides do precipitate species-wide existential threats because those who perpetrate them seek a permanent remaking of the human condition, removing some ways of being human from the earth and installing a new way of being human according to new rules. Even that description fits Bostrom's definition of an existential risk that would lead to a permanent constraint on humanity's potential. As Hannah Arendt detailed, some humans can have their entire political being, their very access to the political, permanently removed from them. To lose one's political being is to lose one's existential condition. "What totalitarian ideologies therefore aim at is not the transformation of the outside world or the revolutionizing transmutation of society, but the transformation of human nature itself. The concentration camps are the laboratories where changes in human nature are tested."[2] There are no permanent, fail-safe solutions—political or technological—to prevent this kind of extreme violence used in transforming human nature by genocidal means. There is no political system to get everyone to agree and eschew violence, and there is also no existence without politics. Attempts to "solve" existential risks once and for all and institute a predictable and controllable definition of the human (in effect abolishing the political) are precisely the origin of genocidal logics. Since humans are inevitably risky toward each other, Arendt adamantly insisted that building durable public institutions, constructing worlds in common, and aspiring to shared long-term public goods are the only way to mitigate the inherent vulnerabilities of the human condition.

In many of his essays on existential risk, Bostrom has often claimed that there is very little academic research being done on this topic. In a 2013 essay, Bostrom remarks that a search of the SCOPUS database delivers 900 papers on dung beetles but the search category of "human extinction" registers fewer than

---

2. Hannah Arendt, *The Origins of Totalitarianism* (New York: Harcourt, Brace, 1973), 458.

50 papers.[3] "It is striking how little academic attention these issues have received compared to other topics that are less important," Bostrom concludes. We did a similar search in March 2020 and found 100 papers on "human extinction" and 2,103 on "dung beetles"—but we also found 7,166 on "genocide" and 138,280 on "extinction" (though these numbers are inflated somewhat since "extinction" is a common term in psychology to denote the cessation of a stimulus, a denotation not relevant to existential risk). In JSTOR, a database primarily of humanities and social science research publications, the search "human extinction" delivers 66,809 results, "genocide" has 43,926 results, and "extinction" has 134,089 results. Similar evidence for the huge amount of research is also found if searching for specific existential risks like nuclear war, pandemics, or genetically engineered bioweapons. Both databases deliver a little over 100 results for "existential risk," but the point is that Bostrom remains too tied to his own selective terms and criteria for what counts as truly existential.

Let's turn the tables here. There is a stunning lack of attention in existential risk studies to the huge amount of research, activism, and human rights work on the history and prevention of genocides. The technocratic outlook and terminological narrowness of Bostrom's assessments are partly at fault, but more disconcerting is the way his work ends up disclosing a colonialist attitude that downplays the history of genocides and Indigenous suffering. There is no evidence that Bostrom (or really any scholar associated with the burgeoning existential risk movement) has informed himself of the multitudinous Indigenous responses to genocide. Bostrom also insists that anthropogenic existential risks only began in 1945, with the invention of nuclear weapons, indicating the first time that humans could wipe out the whole of humanity. The Potawatomi scholar Kyle Powys Whyte points out how the privileging of the current moment

3. Nick Bostrom, "Existential Risk Prevention as Global Priority," *Global Policy* 4, no. 1 (2013): 26.

as truly unprecedented in terms of its potential apocalyptic violence relies on a strategic distancing from past genocidal violence and also implicitly diminishes the knowledge of survival long cultivated by global Indigenous communities. Whyte remarks, "The hardships many non-Indigenous people dread most of the climate crisis are ones that Indigenous peoples have endured already due to different forms of colonialism: ecosystem collapse, species loss, economic crash, drastic relocation, and cultural disintegration."[4] Whyte adds that Indigenous peoples already see themselves as post-apocalyptic and navigating longstanding ongoing crises, connecting post-1945 conditions to the longer history of colonial ultraviolence.

The Terminal Scale

Bostrom's setting the bar for existential risk at apocalyptic levels would make it so that the only event to qualify would be verifiable only after everyone is already gone and no one is left to measure it. This model in effect excludes all existential events in recorded human history, since none of these have been global and terminal. We also find in this argument that taking only the totality of the human species (and perhaps the future transhuman) as the norm of analysis directs the field toward considering as truly existential all-or-nothing visions of sweeping change only at the vastest scales. As Alaimo points out, this abstracted Anthropocene perspective of the human species as a whole works by "scaling up so that human poverty, drought, flooding, or displacement is obscured from sight and the viewer is not implicated."[5] The problems of scale bound up with individual and local group thought and action in connection to global phenomena—the multitiered scale of the existential,

4. Kyle Powys Whyte, "Indigenous Science (Fiction) for the Anthropocene: Ancestral Dystopias and Fantasies of Climate Change Crisis," *Environment and Planning E: Nature and Space* 1, nos. 1–2 (2018): 226.

5. Alaimo, *Exposed*, 153.

embodied subject embedded in a planetary commons—are swept up in this disembodied species-level overview. The medium level of politics, social activism, intersubjective care, and institutional building does factor occasionally into Bostrom's thought, but mostly when he notes that anything done at this level to mitigate existential risk would likely have little effect in stemming most of the dire disasters facing humanity.

In response to Bostrom, most of the scholars in this growing field of existential risk have preferred not to insist exclusively on a standard of total extinction and have steered the discussion toward the broader category of "global catastrophic risks" (and Bostrom himself coedited a book of the same title[6]). Bostrom subsequently has incorporated the broader category of catastrophe into his work but retains his insistence that fully existential risks must be "pan-generational" (affecting all future generations) and go beyond the "endurable" toward the "crushing."[7] He thus remains committed to the view that there are really two main categories of existential risk—human extinction and anything that prevents humanity from reaching a transhuman condition. This position still remains prominent in the field of existential risk, as evidenced, for example, in Phil Torres's definition of existential risk as "any future event that permanently prevents us from exploiting a large portion of our cosmic endowment of negentropy to create astronomical amounts of those things that we find valuable."[8] Toby Ord also follows along similar lines, including the supposed "plateauing" of humanity as an existential failure in his central definition: "An existential risk is a risk that threatens the destruction of humanity's longterm potential."[9]

---

6. Nick Bostrom and Milan M. Ćirković, eds., *Global Catastrophic Risks* (Oxford: Oxford University Press, 2008).

7. Bostrom, "Existential Risk Prevention," 17.

8. Phil Torres, "Facing Disaster: The Great Challenges Framework," *Foresight* 21, no. 1 (2019): 5.

9. Toby Ord, *The Precipice: Existential Risk and the Future of Humanity* (New York: Hachette, 2020), 70.

As we will see, the most common effect of this definition is to convince that glorious posthuman futures are not to be seen as merely possible but rather obligatory—we *must* make good on these astronomical long-term potentials, anything else is a catastrophic failure. Fear of disappointing this big Other, the imaginary great future of humanity, runs rampant across the field. Such imperatives make it possible to understand transhumanism in the context of messianic history—especially Gnosticism and the Christian narrative of redemption.[10] Though it is not our focus in this book, the ostensibly secular and scientific arguments of transhumanism, especially when they push for the disembodiment of mind into AI, defer to deep-seated patterns of religious thought.[11]

Typological Thought

The bulk of Bostrom's 2002 essay is devoted to the classification of existential risks. He launches into this project by proposing four idiomatic classificatory categories: "Bangs" (a sudden extinction), "Crunches" (human civilization is stunted and humans never achieve transhumanity), "Shrieks" (only a narrow transhumanity is attained), and "Whimpers" (some transhumanity is achieved but it also impoverishes things we value). These remarkable terms are inspired by T. S. Eliot's "The Hollow Men" (1925). We are left

10. Several proponents of transhumanism have welcomed this religious eschatological alignment. See Calvin Mercer and Tracy Trothen, eds., *Religion and Transhumanism: The Unknown Future of Human Enhancement* (Santa Barbara, Calif.: Praeger, 2014).

11. See Erik Davis, *Techgnosis: Myth, Magic, and Mysticism in the Age of Information* (Berkeley, Calif.: North Atlantic Books, 2015). This is also Jason Lanier's point in his take on the issue of AI as destroying or saving humanity: "This is not an honest conversation. . . . People think it is about technology, but it is really about religion, people turning to metaphysics to cope with the human condition." Raffi Khatchadourian, "The Doomsday Invention," *New Yorker*, November 23, 2015, https://www.newyorker.com /magazine/2015/11/23/doomsday-invention-artificial-intelligence-nick -bostrom.

to wonder if these categories are forged by philosophical fiat, or by scientific consensus, or posited as metaphorical guidelines. Yet how should one construe the overlap of the methodological contributions of science and the literary imagination? We have no problem with employing metaphorical speculations to aid in thinking extinction, but why these metaphors? Are they the only categories plausible or available?

Before we look closer at Bostrom's list of existential risks, this is the right moment to pause and reflect on the kind of philosophical style and its effects on argumentation that Bostrom most often employs in this essay and many subsequent others. While his philosophical principles are primarily drawn from utilitarian rationalism, risk management, and speculative transhumanism, he generally organizes his essays by providing typologies and lists, a writing style that has its own philosophical effects. The main advantage of this presentation style is that he delivers an overview of existential risks with clarity and concision. Readers can quickly assess the looming existential threats at a glance, and it feels like one is reading a kind of high-level governmental briefing. The disadvantage is that this presentation style can shortcut further philosophical analysis and critique. The brevity and directness of lists and typologies also means that many of his philosophical assumptions proceed unmarked and unargued. What *kind* of philosophy is typological thinking? Typological lists offer little if any forum for discussion and extended reflection. Typological thought tends to dispense with any extensive citation or broader background reading. It suggests that the issues already have been distilled to bullet points to summarize what we need to know in order to cognize existential risks.

Actually, Bostrom cites very few philosophers from any stripe or school in these essays. Interpreting this absence, it would seem that Bostrom does not find that traditional philosophy has much to offer in extreme existential situations. Is philosophy too human, too discursive, too argumentative, too slow, too powerless in the face of tremendous technological developments and planetary catastrophes? Bostrom does not dwell on the problems of philosophy itself.

Yet his philosophical writing is largely a rejection of the philosophical discipline, an admission that philosophy has not done enough to make the world better or safer, and that it would be more efficient and effective to "align" philosophy with technologies to do the work that philosophy alone had tried but failed to accomplish. Instead of philosophy realizing a world of enlightened reason, superintelligent technology will have to do it.

Typological thought slants analysis away from much of the humanistic toolkit of philosophy and toward top-down policy propositions and heroic technocrats, implying that these are the true agents who will protect humanity from the disastrous future. Such typologies tend to be decontextualized and ahistorical pronouncements that rush the reader toward immediate judgment—"Here is a list of doomsday scenarios, there's no time for debate or dallying, we must act now!" The existential risks that Bostrom lays out are serious and we do not mean to belittle Bostrom's clear commitment to encourage further research from a wide variety of approaches to analyzing existential risk. We are concerned rather with how Bostrom's own stylistic reductiveness shapes his argument as well as the trend of argumentation across the field.

Finally, on the topic of typological philosophy, Bostrom's philosophical style does breed a tremendous amount of jargon. The language of bangs, crunches, shrieks, and whimpers is a clear example,[12] but even more basic terms function as jargon. Terms such as "existential," "risk," and "global public goods" (ER, 4), along with categories of existential risks themselves such as "dysgenic pressures" (ER, 11) and "technological arrest" (ER, 12), become

12. Nicole Shukin points out that these terms describe conditions of extreme violence and mass suffering, but here are used in a conspicuous display of nonchalance. Shukin adds that Bostrom's "politics of affect" and dispassionate prose cultivates an unsentimental, hardened, and abstracted attitude toward existential risk. Nicole Shukin, "Prospecting Future Ruins: On the Speculative Character of Existential Risk," paper presented at Wrack Zone: Association for Literature, Environment, and Culture in Canada conference, University of Victoria, June 23, 2018.

expedient insider terms used without reflective analysis. To be sure, all philosophies have jargon. There is nothing wrong with using specialized terms. We employ plenty of jargon in this book—we're not shy about it. Our point here is that jargon can be mitigated with a commitment to building self-critique and a process of social mediation and public responsibility into the methodology of one's analysis. Marginalizing these practices of critique, sharing, and consultation leads Bostrom to confirm his own view that these longstanding humanist modes of inquiry used to build the public sphere will be of little use in confronting the overwhelming violence of existential risks when they do occur.

Even disasters breed jargon. One prescient example of the brutal effect of this can be found in Naomi Klein's analysis of disaster capitalism and the mobilization of neoliberal textbook language into schemes to transform national public services into open markets in *The Shock Doctrine*.[13] Bostrom's "apocalypse now" methodology has little interest in examining how the effects of catastrophes are unevenly distributed, exacerbating existing social inequalities, and further entrenching economic disparities and dispossession among peoples and between nations.

It bears recalling here Theodor Adorno's trenchant study of what he called the "jargon of authenticity"—a critique Adorno specifically launched against the proliferation of jargon in existential thought, especially in Martin Heidegger's version. Adorno points to how quickly even the term "humanity" can become jargon when its invocation is used to deflect from any specific devotion to social betterment. When "humanity" is transformed into jargon, "Humanity becomes the most general and empty form of privilege."[14] While exclaiming this jargon, the human domination of other humans and nature continues unabated. The jargon of

13. Naomi Klein, *The Shock Doctrine: The Rise of Disaster Capitalism* (New York: Random House, 2007).

14. Theodor Adorno, *The Jargon of Authenticity*, trans. Knut Tarnowski and Frederic Will (London: Routledge, 2007), 53.

humanity, Adorno adds, "caricatures the equal rights of everything which bears a human face, since it hides from men the unalleviated discriminations of societal power: the differences between hunger and overabundance. . . . In the mask of jargon any self-interested action can give itself the air of public interest, of service to Man. Thus, nothing is done in any serious fashion to alleviate man's suffering and need" (54).

One example of Bostrom turning "humanity" into this type of jargon comes in a 2013 essay where he supplies a graph of the rise of world population over the last century showing a slow, consistent rise of total population from 1900 to 1950, and a steeper rise (called the "great acceleration") from 1950 to 2010. Bostrom provides the following caption as interpretation: "Calamities such as the Spanish flu pandemic, two world wars, and the Holocaust scarcely register. (If one stares hard at the graph, one can perhaps just barely make out a slightly temporary reduction in the rate of growth of the world population during these events)."[15] The upward graph is taken as "proof" that these "calamities" were not really existential events. The generic category of humanity, now turned into a population data point, has survived and moved on. No other lessons are gleaned here for existential risk. Bostrom, of course, is not denying historic suffering; the point is that the graph is used to confirm Bostrom's own methodology, concealing his controversial polemic about what truly constitutes an existential risk as simply "confirmed" by the data.

Beginning with the End

Turning back to Bostrom's list of "bangs," these potentially devastating events are the core of what is commonly understood as the extinction crisis facing humans. His category of "bangs" lists the following in descending order of probability (beginning with what

---

15. Bostrom, "Existential Risk Prevention," 18.

Bostrom estimates to be most probable, although he assigns no specific risk calculations here): deliberate misuse of nanotechnology, nuclear holocaust, we are in a simulation (a cosmic algorithmic program run by some form of superintelligence) and it gets shut down, a superintelligence of our own creation leads to our undoing, a genetically engineered bioagent, an unforeseen physics disaster of our own creation (for example, a new particle that annihilates Earth into a vacuum), a naturally occurring pandemic, an asteroid impact, and runaway global warming (as likely occurred on Venus). Such lists exemplify the "rhetoric of probability" we discuss in chapter 2, where existential risk discourse takes these probabilities as a spur for speculation on dystopian or utopian outcomes rather than actually carrying out calculations. With hindsight of the coronavirus in 2020, it's curious that pandemics and extreme global warming rank lowest in probability, although, as we know, Bostrom would not rank the current concerns as existential, merely catastrophic.

In compiling this "worst ever" list, Bostrom often proposes a utopian counterpoint to each dystopian scenario. The deliberate misuse of nanotechnology follows along the same route as the utopian possibilities of this technology. Things could go very badly in many ways, but, "If things go well, we may one day run up against fundamental physical limits. . . . But here we are talking astronomical time scales" (ER, 14). Extinction looms, but so do the most immense successes imaginable. Indeed, for Bostrom, the failure to eventually achieve the utopia of "astronomical" value in the long run and leave the human condition behind will always be dystopian. Bostrom keeps insisting that it would be an existential risk *to not pursue* utopia (a position we will scrutinize further below), by limiting ourselves to an "extremely narrow band of what is possible and desirable" (ER, 5). This oscillation between dystopia/utopia continually appears in Bostrom's argument. It parallels his seesaw of claims that we have various biases that tend to underestimate some existential risks and overestimate others, suggesting we will always undershoot or overshoot a problem.

To his credit, Bostrom's foundational 2002 essay intriguingly fosters an openness toward science fiction–like philosophical "what if" scenarios of planetary catastrophe. In another essay drafted just prior to the "Existential Risks" essay, Bostrom walked through some speculative arguments about whether or not it might be the case that humans are living in a computer simulation run by an exponentially higher intelligence.[16] If we are living in virtual cosmos, it would constitute our ultimate existential risk since the simulation could be shut down at any time. Extinction would have an entirely different meaning than the biological definition we assume right now. This provocation that extinction might be something radically different raises other far-reaching problems. If life is substrate-neutral and can appear in different underlying mediums, materials, and programs—or if life is implicit in all matter (panvitalism) like mind is for panpsychists—then extinction and existential risk will require massive redefinition, pushing the concepts of extinction and risk beyond coherence.[17] Methodologically, Bostrom's range of hypothetical existential risk thinking has fostered an open-minded inquiry into any conceivable end of the world. However, as we discuss further in chapter 2, the philosophers of existential risk do not want their work confusing science and science fiction, and rationalist utilitarian and probabilistic thinking does not methodologically align with speculative metaphysics (or existential thinking, for that matter). The field requires a coherence to the definition of existential risk and extinction, but these methods of reasoning and calculating probability run into a number of inconsistencies and incoherencies.

Bostrom's other categories of "crunches," "shrieks," and "whimpers" carry on in the same mode, mixing scientifically evident prob-

16. Nick Bostrom, "Are You Living in a Computer Simulation?," *Philosophical Quarterly* 53, no. 211 (2003): 243–55.

17. For a further discussion of the problems of incoherence regarding definitions of extinction in vitalist and eliminative materialist views of life, see Joshua Schuster, "Life after Extinction," *Parrhesia* 27 (2017): 88–115.

lems with hypothetical scenarios to classify near-term planetary disasters and long-term speculative social, biological, and techno-logical bottleneck phenomena (the notion that we have to squeeze through a very risky historical period to get to a more stable and advanced society). Since Bostrom is supplying a brief bird's-eye view of existential risks, he does not bother to incorporate much in the way of a critique of existing social inequalities. While re-source depletion and ecological destruction do rank as possible "crunches," the unexamined economic and social arrangements that underlie these risks include capitalism, colonialism, and any eco-nomic system that requires endless growth to function. Moreover, we find no analysis of the fact that the pursuit of new technologies combined with repressive political regimes will exacerbate the suffering of those who already find their lives exposed to a high degree of existential risk across the planet. Kathryn Yusoff remarks that Anthropocene discourse of the generic human species ends up effacing how material infrastructures and resource extraction follow on the same tracks of colonialism and racialized power. The super-technologies and superintelligence that will save "us" will have to run on the minerals and extractive resources taken predom-inantly across the globe from Brown and Black communities and nations that historically have been treated as existentially expend-able. Yusoff writes, "To be included in the 'we' of the Anthropocene is to be silenced by a claim of universalism that fails to notice its subjugations."[18] While some are existentially secure—and can raise inquiry into how such security is possible—others have long been exposed to everyday existential insecurity. Bostrom's risk scenarios have no specific insight into how this structurally uneven distribu-tion of existential risks also produces existential privileges.

Bostrom's expanded range of "what if" scenarios that cover im-mediate concerns as well as hypothetical long-range possibilities

18. Kathryn Yusoff, *A Billion Black Anthropocenes or None* (Minneapolis: University of Minnesota Press, 2018), 19.

can make it harder to assess and prioritize more urgent and likely calamities. For example, while he mentions near-term concerns over environmental crises, they are not particularly highlighted or examined in the context of any broader environmentalist commitments. Bostrom points to the risk of greenhouse gases on Earth leading to runaway global warming (ER, 10) such as found on Venus—an ecological problem, to be sure, but one pitched at the level of an improbable total destruction of habitable life on a cooked Earth and not particularly engaged with the everyday work of ecological care. Another environmental existential risk mentioned is the depletion of primary resources (nonrenewable fuels, timber, water) leading to civilizational collapse. These limited resources are certainly immediate concerns, but Bostrom and others in the existential risk field who tend to rely on typological lists of disaster scenarios have not prioritized the nonspectacular, hard work of researching water protection or divestment from oil and coal and the collaborative energy transition to come. Bostrom tends to keep his eyes peeled for quick and dramatic risks. He offers only occasional nods to what Rob Nixon calls the "slow violence" of accumulating environmental degradations that have become features of daily life for the global poor.[19] Because existential risk examines only the extremes of human suffering, there is little interest in small-scale fixes.

Here we can point out that all of Bostrom's existential risks focus strictly on humans and have nothing to say regarding extinction threats to animals. Bostrom does not show much interest in animal life beyond a rudimentary utilitarian concern to reduce suffering across species. So far, none in the field of existential risk seem to have applied the analysis of human existential risks with regard to animals or studied how the history of animal extinction informs human extinctions. Winona LaDuke shows that the continued elimination of Indigenous peoples intersects with the ongoing destruc-

---

19. Rob Nixon, *Slow Violence and the Environmentalism of the Poor* (Cambridge, Mass.: Harvard University Press, 2013).

tion of animal and plant species: "There have been more species lost in the past 150 years than since the Ice Age. During the same time, Indigenous peoples have been disappearing from the face of the earth. Over 2,000 nations of Indigenous peoples have gone extinct in the western hemisphere, and one nation disappears from the Amazon rainforest every year. There is a direct relationship between the loss of cultural diversity and the loss of biodiversity."[20] The diversity of peoples and languages mirrors the diversity of species and the disappearance of peoples historically has coincided with the disappearance of the planet's biodiverse life.

Doom Patrol

Bostrom's essay details some reasons why existential risks pose a unique set of challenges for human responsibility and traditional moral reasoning. He suggests that the magnitude of risk is so large that conventional use of trial-and-error to correct problems is insufficient. For some existential risks, especially new ultrapowerful technologies and weapons, we may only have one chance to get it right. The first person or group to achieve such technologies and weapons may have the power to dominate or destroy everyone. Everything depends on being able to shift what might be an existential risk into, if not full prevention, a "mere" catastrophe that would allow us time to have more than one chance to act—time to learn and teach ourselves how to live together again. Bostrom's estimation that we only get one chance can also be questioned as it assumes a fully formed AI or doomsday weapon functional as if overnight. Bostrom argues that there may be situations where some kind of "preventive action" (ER, 3) is needed, a potentially militaristic claim akin to justifications for "pre-emptive" war-making. Bostrom also airs the concern that any current institutions, state actors, or moral norms would be insufficient to the task of avoiding

20. Winona LaDuke, *All Our Relations: Native Struggles for Land and Life* (Cambridge, Mass.: South End Press, 1999), 1.

or managing existential risks. The lack of precedent, the likelihood of panic, and the probability that we underestimate such violence and overestimate the time and skill we have in solving terminal planetary problems are all factors suggesting current norms will not abide at existential threat levels. A truly existential event would induce a radical "state of exception," one in which there may even no longer be a state or identifiable political authority.

Relying on human capacities for responsibility in exceptional events presents other risks, and many ethicists argue that a more scientific and "objective" approach to ethics may involve taking ethics out of human hands, favoring institutional or technological decision-making. Existential risk scholar Phil Torres rightly emphasizes vigilance toward what he calls "agent-tool couplings,"[21] situations in which powerful tools can fall into the hands of bad agents. Among bad agents, Torres lists dictators, misanthropists, and believers in apocalyptic violence as redemptive according to eschatological religious narratives. Torres also adds "ecoterrorists" and other "misguided ethicists" who might want to erase humans from the Earth. However, how is this vigilance able to make distinctions between real people who strive to eradicate all of humanity on a moment's notice and discursive statements and performances of the apocalyptic imagination, however nihilist or ecofascist? In any case, one massive problem with Torres's list is that any agent with a grudge against "humanity" becomes a risk that supposedly justifies using "preventive action" to subdue. Yet agents who use such preventive action also pose new risks of their own to the planet by calling for mass surveillance and incarceration as the only apparent way to weed out malicious agents before they strike. A situation could arise in which there are multiple agents claiming different kinds of emergencies that prioritize different kinds of existential risks and use extrajudicial and preemptive methods to destroy what they deem as nefarious agent–tool couplings. The violent history

21. Phil Torres, *Morality, Foresight, and Human Flourishing: An Introduction to Existential Risks* (Durham, N.C.: Pitchstone, 2017), 95.

of this kind of totalitarian paranoid power is obvious and the existential risk theorists who find themselves airing these scenarios as unsavory but perhaps necessary need to answer to why their methodology offers little else in terms of "foresight."

Despite these misgivings and doubts on institutional resilience and individual responsibility, many of the researchers involved in this field have concentrated their advocacy efforts on global governance and scientific collaboration. Bostrom and others continue to support notions of sustainable good governance in ecological, technological, and economic systems, though these systems may not always be compatible and may still fail in an extreme event. Although the need for planetary governance would seem to point to a discussion of collectivism or planetary political commons, researchers in existential risk tend to look askance at any form of radical political change or unified social horizon. They are not calling for a movement to defund the military even as they advocate for nuclear disarmament, although the contribution such funding makes to the proliferation of weapons surely constitutes an existential risk (and yet, military programs like DARPA are major funders of superintelligence research). They are not leading researchers into the recent waves of white nationalism and authoritarianism, movements that have articulated a vision of remaking the planet through violence, xenophobia, and oppression (and here it bears mentioning that a number of libertarian transhumanists have quietly supported attitudes of white escapism).

Perhaps most conspicuously, the field of existential risk offers no substantial critique of capitalism, in which an indefinite ongoing demand for growth and resources poses an eventual existential crisis (historically, economic risk and capitalism are synonyms indicative of the spread of financial risk and reward across social classes[22]). Existential risk researchers seem to prefer neoliberal practices setting the current standards of value according to markets

22. See Beck, *Risk Society*; Niklas Luhmann, *Risk: A Sociological Perspective*, trans. Rhodes Barrett (New York: Routledge, 2017).

regulated lightly along utilitarian lines (interest in universal basic income is not touted as a way to abolish exploitative capitalism and extractivism). In *Risk Criticism,* Molly Wallace describes how the "outsourcing of risk" has been built into capitalism in which nations and corporations "play the unevenness of the global market by capitalizing on the gaps in risk regulation."[23] Often the worst realities of these risks are transferred onto disenfranchised peoples who have a long and brutal road to haul for redress. Truly dampening the recurrent recklessness of economic risk would have to involve transitioning out of capitalism. But it seems to be easier to imagine changing the existential structure of humanity than to imagine the end of capitalism. Indeed, capitalism now looks to existentiality as its new frontier, taking on the core list of traditional existential concerns—freedom, moods, intentionality, natality, and mortality—as markets for engineering and programming.

Endgame Moves

Bostrom closes his initial essay with a "rule of thumb for moral action" that he dubs "maxipok": "Maximize the probability of an okay outcome, where an 'okay outcome' is any outcome that avoids existential disaster" (ER, 25). Hardly stirring, this instruction is vague and directionless, closer to the Silicon Valley dictum of "don't be evil" than a clearly defined new categorical imperative. But the jargonistic phrasing of "maxipok" proves to be deceptive, since Bostrom makes it clear that he does not condone just any outcome for humanity. In contrast to this seemingly nonchalant expectation for an "okay outcome," Bostrom insists in the essay that "technological arrest," whereby humans never "transition to the posthuman world" (ER, 12), would constitute its own kind of existential failure. To just be okay is not okay. Built into Bostrom's philosophy is the

---

23. Wallace, *Risk Criticism,* 66-67.

argument that to remain existentially the same (merely "okay") and not achieve a posthuman triumph would be an ultimate failure.

The audacity of arguing that merely remaining human would be a kind of existential threat boggles the human mind. In an essay drafted in 2007, Bostrom outlines four primary possible human futures: extinction, recurrent collapse, plateau, and posthumanity.[24] In Bostrom's thinking, only the last category would be a worthy objective. The essay defines "recurrent collapses" as cycles of violence and assorted disasters in which civilization is set backward, advances, and is set back again. "Plateauing" indicates humanity failing to keep up the tremendous rate of technological advances and health improvements brought about in the past few centuries. Humans would plateau if they are not able to live substantially longer lives, eradicate most causes of suffering, and develop cognitive improvements either biologically or artificially that are at least two standard deviations above current human maximums (ER, 20). The timeline for escaping the human plateau is less important than its eventual accomplishment, which for Bostrom constitutes the ultimate definition of the good.

If humans don't commit in favor of pursuing technological greatness in the short or even long term, is this really "plateauing"? Picasso is said to have uttered upon seeing the Paleolithic-era cave drawings in Lascaux, "We have invented nothing." These images still stir because we don't dismiss this art as an order of aesthetic magnitude or "plateau" below us. Aesthetics and ethics facilitate capacities that are cognitively enjoyable as well as enhancing; technological development is not the only barometer for expanded consciousness and experience. And it's not as though we've explored the potential of what a much more widespread aesthetic and environmental education would do to society. Artificial intelligence or posthumanism might bring about a diminishment of experience in

24. Nick Bostrom, "The Future of Humanity," in *New Waves in Philosophy of Technology*, ed. Jan-Kyrre Berg Olsen, Evan Selinger, and Søren Riis (New York: Palgrave McMillan, 2009).

some ways, as computers process more of the world with or without us. There is also ongoing evolution and disruption in a plateau, and more to come in Deleuze and Guattari's "a thousand plateaus"! The "plateau" is a metaphor that points to other dynamic metaphors of creative commons—consider then what futures for humanity arise by foregrounding the unruly garden, spaceship Earth, or another figure for the existential planetary journey?

2. Probability and Speculation

AS AN OFFSHOOT of analytic philosophy, existential risk is an inter-disciplinary field that boasts of many actual and desired ties to the natural sciences. For existential risk analysts, "science" means three closely related things. First, existential risk claims to be a rational and sometimes empirical, data-driven approach to theorizing and mitigating the possibility of human extinction. In this case, *rational* means quantification and mathematical calculation, a combination of probability theory (used in risk analysis by the insurance industry, for example) and utilitarian ethics. Second, existential risk scenarios are meant not to be fantastical but plausible from the point of view of current knowledge. In this case, science means *a realism of plausible futures extrapolated from current knowledge.* As we show here, this distinction echoes arguments in literary studies for the need to distinguish speculative fiction from science fiction. Third, in its effort to be scientific, existential risk is saturated by the tacit influence of science fiction, a genre of literature intricately bound up with the history of science. With these three meanings of "science," we are concerned with how science fiction imaginaries morph into a science of the possible.

For us, this triple relation with science means that existential risk can and should be critiqued from the vantage of science and technology studies, where scholars from N. Katherine Hayles to John Johnston and Ruha Benjamin already address transhumanism and AI.[1] In this chapter, our second critique of existential risk, our critical process employs techniques of second-order observation. One system, science and technology studies, will observe the observations of another, existential risk, in order to understand its conditions of possibility and its blind spots. Second-order observation includes analyses that are both internal and external to the field being observed, resituating the terms immanent to the field and using these terms in a self-critical way. At stake is how we interpret the eschatological implications, the very *outer frame,* of "our" ostensibly secular scientific worldview.

The theoretical models used by existential risk analysts vacillate among deep time speculations, unknown risk horizons, and policy making in the present. What kind of scientific reasoning best fits these timescales and the "unprecedented" nature of the events in question? The philosophers of existential risk deploy a model of probability far more than actual calculation. But they promise that their account of "possible" extinctions and the actions that will help humans avoid them can and will have a quantitative foundation. Their claim is that *low probability* plus *extinction-level significance* means we should care more about existential risks than anything else—and that this level of analysis is truly the ultimate priority of "effective altruism." In the case of avoiding remotely probable existential risks, the benefits and rewards are postponed for a future distant enough that it will be hard for most to care deeply about it, with the added complication that many of these extinction scenarios might not be possible in the first place. It would not be fair to say that existential risk is fundamentally incoherent when assessing risk probability and leave it at that. But this caricature gets at something that we will take up with more rigor in this chapter.

1. N. Katherine Hayles, *How We Became Posthuman: Virtual Bodies in Cybernetics, Literature, and Informatics* (Chicago: University of Chicago Press, 1999); John Johnston, *The Allure of Machinic Life* (Cambridge, Mass.: MIT Press, 2008); Ruha Benjamin, *Race after Technology: Abolitionist Tools for the New Jim Code* (Malden, Mass.: Polity, 2019).

The field of existential risk combines utilitarianism and probabilistic risk analysis toward a fortuitous "ok" outcome (Bostrom's "maxipok"). On the one hand, existential risk analysts aim at the utilitarian goal of establishing the greatest good for the greatest number. On the other hand, they seek to use mathematical probability models to estimate the chance that a given cause will lead to the extinction of humanity, or even of life in general.[2] The crux of existential risk analysis is the synthesis of these two modes of calculative reasoning. But this crux is also a failure: as we show in this section, existential risk analysis is *constituted* by a radical mismatch between method and object of study. This mismatch leads to a comic effect: the wonderfully strange spectacle of serious philosophers (Bostrom), scientists (Sir Martin Rees), and entrepreneurs (Elon Musk) discussing how to mitigate the risk of deep future extinction events as though they were calculating the probability that one will die in a car accident based on ample statistics about the frequency of such events in the past. The effect is similar to that of Michael Madsen's documentary *Into Eternity,* when we watch engineers become speculative philosophers as they ponder the one hundred thousand-year time frame of their deep geological repository for storing nuclear waste, especially when Madsen asks them what human societies might be like in such a distant future.[3] What Mark McGurl calls "posthuman comedy" is here the effect of rational solutions applied to cover over

2. Classical probability models that would address the odds, for example, of rolling a six with dice cannot be used to study existential risk. Most often, the probability models used are conditional or Bayesian, though as we argue below, this is often unstated or fuzzy, at least for the nonexpert. Conditional probability is the probability of A given that condition B has already happened. Bayesian probability is defined below. See Ian Hacking, *Introduction to Probability and Induction* (Cambridge: Cambridge University Press, 2001); and for the history of probability theory, Ian Hacking, *The Taming of Chance* (Cambridge: Cambridge University Press, 1975).

3. Michael Madsen, dir., *Into Eternity* (Copenhagen: Films Transit International, 2010).

the unthinkable, unimaginable, and uncontrollable otherness of any future that stretches beyond a few human generations.[4]

Existential risk veers from "comic" analyses of extremely remote scenarios, such as the "aestivation hypothesis" (aliens are sleeping so as to wait billions of years for the universe to cool down enough to run cosmic-size computers), to "tragic" assessments of the high near-term likelihood of some massive extinction-inducing event that would be predictable but not avoidable.[5] If we were to start calculating the probability that a climate tipping point or AI will bring an end to our evolutionary line, then not only would we be unable to arrive at a reliable estimate that bears any analogy with, for example, finance or insurance risk analysis, we wouldn't even know if the event is possible in the first place. We would be trying to estimate the probability of something that can only happen once, without any evidence to go on. Not only that, we would be trying to use this knowledge to act in a way that will prevent it from happening. Or at least, because existential risk analysis operates within a fundamentally probabilistic universe, the aim is to lower the fatal event's probability: the closer to zero, the more immune humanity's potential will have been.

For some, this might already be enough to dismiss the idea of rational study of remote existential risk scenarios. As one AI researcher reports, "I don't worry about [AI induced extinction] for the same reason I don't worry about overpopulation on Mars."[6] There is a common-sense idea that we should not spend too much time or too many resources to prevent something when we don't know if it's possible (though Bostrom and others would remind us that common sense is what prevents us from seeing the implications

    4. Mark McGurl, "The Posthuman Comedy," *Critical Inquiry* 38, no. 3 (2012): 533–53.
    5. Anders Sandberg, Stuart Armstrong, and Milan Ćirković, "That Is Not Dead Which Can Eternal Lie: The Aestivation Hypothesis for Resolving Fermi's Paradox," (self-published, May 10, 2017), https://arxiv.org/pdf/1705.03394.pdf.
    6. Khatchadourian, "Doomsday Invention."

of probability theory clearly enough to act). As our characterization of existential risk's model of probability and utilitarianism suggests, we lean toward skepticism about the field's claim to rationality and scientific rigor. For readers steeped in critical theory, such a quantitative approach to politics, ethics, and extinction may look to be another chapter of the dialectic of Enlightenment, bound to end badly. But the epistemological assumptions of existential risk should attract greater attention and critique—not dismissal as another overreach of rationalism, but analysis and historicization.

The Rhetoric of Probability

Writing for *Vox,* journalist Dylan Matthews covered the Effective Altruism Global Conference at the Google campus in Mountain View, California, in the summer of 2015. The title of his article—"I Spent a Weekend at Google Talking to Nerds about Charity. I Came Away . . . Worried"—divulges his take on the ideas and affects that circulated at the meeting.[7] "Effective altruism" is the philanthropic practice of attempting to do the greatest good through means that are efficient and data-driven rather than sentimental. Many in the existential risk community also subscribe to the effective altruism movement, such as philosopher Toby Ord, founder of the society Giving What We Can (a group admirably committed to donating at least 10 percent of income). In Oxford, Bostrom's Future of Humanity Institute shares office space with the Centre for Effective Altruism. Effective altruists see themselves embracing "the cold, hard data necessary to prove what actually does good."[8] This makes the movement a candidate for study from the perspective of science and technology studies, as an ethical philosophy that claims for itself scientific imprimatur. Matthews says that he

7. Dylan Matthews, "I Spent a Weekend at Google Talking to Nerds about Charity. I Came Away . . . Worried," *Vox,* August 10, 2015, https://www.vox.com/2015/8/10/9124145/effective-altruism-global-ai.

8. Matthews.

identifies as an effective altruist; he also admits that "EA is very white, very male, and dominated by tech industry workers."[9] The topic of existential risk occupied center stage at the conference in 2015, and it gave him pause even as someone who embraces quantitative ethics.

Matthews's skeptical account of this "X-risk" takeover is a good example of how existential risk combines fantastical, deep-time probability calculations with utilitarian ethics. In the example he cites, a panel featuring Bostrom and Musk, the starting point for determining the greatest good for the greatest number of persons was to calculate the greatest number of lives. The presenters did so not with respect to a narrow time frame, as in the greatest number alive on Earth today or during a 100-year period. Instead, they reasoned that if humanity lasts "another 50 million years," then "the total number of humans who will ever live is . . . 3 quadrillion."[10] But they went on to decide that this number fails to take into account future extraterrestrial inhabitants of the solar system, "the potential value of our posthuman future," or what Phil Torres calls the "astronomical value thesis."[11] Given the same arbitrary time scale of 50 million years, they concluded that the number of people we need to take into account is more like 10^{52} lives of 100 years each—a vastly greater number than 3 quadrillion, so much greater that the mathematical notation for exponents seems more accessible than obscure words like *sexdecillion*.

Bostrom then shifts from such uncountable numbers to discussing the ethical implications for us today: "Even if we give this 10^{54} estimate 'a mere 1% chance of being correct,' . . . we find that the expected value of reducing existential risk by a mere *one billionth of one billionth of one percentage point* is worth a hundred billion times as much as a billion human lives."[12] As Matthews continues to paraphrase,

9. Matthews.
10. Matthews.
11. Torres, *Morality, Foresight, and Human Flourishing,* 41.
12. Matthews, "I Spent a Weekend at Google."

the number of future humans who will never exist if humans go extinct is so great that reducing the risk of extinction by 0.00000000000000001 percent can be expected to save 100 billion more lives than, say, preventing the genocide of 1 billion people. That argues, in the judgment of Bostrom and others, for prioritizing efforts to prevent human extinction above other endeavors. This is what X-risk obsessives mean when they claim ending world poverty would be a "rounding error."[13]

Since we read this article, we have mentioned Matthews's rather satirical report to a number of friends and colleagues. Their reaction is always amusement at the comical absurdity of this scenario. They are right, in a sense, but in this case the reductio ad absurdum of deep-time utilitarianism is strange because it is also perfectly rational, if "rational" means consistent with the concepts thinkers like Bostrom have applied. Notwithstanding their highly speculative assumptions about the time scale of fifty million years and interplanetary travel, the numbers reported by Matthews are faithfully deduced by combining probability theory with the first principle of the mathematical ethics of utilitarianism, even if the greatest "good" is reduced to mere existence for the greatest number. This is the "effective" side of effective altruism, now stretched to an unimaginable future.

In another example, Bostrom offers a box essay in *Superintelligence* about the human lives and happiness that are really at stake when it comes to the risk of malevolent AI. Different from Matthews's example, he adds a quantification of happiness that goes beyond the negative value of avoiding the foreclosure of countless future lives:

Assuming that the observable universe is void of extraterrestrial civilizations, then what hangs in the balance is at least 10,000,000,00 0,000,000,000,000,000,000,000,000,000,000,000,000,000,000,000 ,000 human lives. . . . If we represent all the happiness experienced during one entire such life with a single teardrop of joy, then the happiness of these souls could fill and refill the Earth's oceans every

13. Matthews.

second, and keep doing so for a hundred billion billion millennia. *It is really important that we make sure these truly are tears of joy.*[14]

This implied utilitarian calculus stretches the limits, to put it mildly, of any known means of deciding how we should act, what is right and wrong, and how to calculate the odds of a given extinction. Yet they remain calculations. They have a certain rationality despite the absurd ambition. They also provide the foundations for a normative prescription for action, even if it reads slightly tongue-in-cheek given the image of an ocean filled with tears of joy. The aspiration here is to make an ethics of extinction avoidance rational by grounding it in the relation between probability of a given risk and the idea that doing the right thing means dividing the total amount of total happiness, now and in the future, by the total number of humans. Working with such vast numbers is what gives the idea that even *slightly* lowering the probability of an extinction event is worth "astronomically" more than any justice achieved in the present.

Another example of Bostrom's use of probability appears in the first chapter of *Superintelligence,* where he posits that AI will reach human-level intelligence during our century. In this case, the evidence with which to calculate such probabilities comes secondhand from surveys of AI experts recalibrated by Bostrom. Rather than probabilities calculated on the basis of known conditions such as the two sides of a coin or all the data about car accidents used by insurance companies, we have researchers guessing at probabilities, other researchers averaging them out, and Bostrom reporting the results. So he is right to admit that

> small sample sizes, selection biases, and—above all—the inherent unreliability of the subjective opinions elicited mean that one should not read too much into these expert surveys and interviews. They do not let us draw any strong conclusion. But they do hint at a weak conclusion. They suggest that (at least in lieu of better data or analysis) it

14. Nick Bostrom, *Superintelligence: Paths, Dangers, Strategies* (Oxford: Oxford University Press, 2014), 103.

may be reasonable to believe that human-level machine intelligence has a fairly sizeable chance of being developed by mid-century, and that it has a non-trivial chance of being developed considerably sooner or much later; that it might perhaps fairly soon thereafter result in superintelligence; and that a wide range of outcomes may have a significant chance of occurring, including extremely good outcomes and outcomes that are as bad as human extinction. At the very least, they suggest that the topic is worth a closer look.[15]

Bostrom rhetorically hedges several layers of uncertainty, writing not just that we should take the numbers with a grain of salt, but that a "weak conclusion" "suggests" that it "may be reasonable" to believe in a "fairly sizeable" chance of the development of human-level AI, "which might perhaps fairly soon" become superintelligent, which could then mean a "significant chance" of human extinction alongside the more neutral and optimistic possibilities. This rhetoric of probability does not amount to anything very different from an unguided estimate. More generously, we can say that the changing opinions of experts allow for guesses that are easier to trust secondhand, and that these guesses can be updated as new evidence arrives. Yet they remain guesses about several things that we do not know to be possible in the first place.

There would seem to be a qualitative difference between such "probabilities" (if this is still the right term) that involve educated guesses and secondhand surveys, and the kind of probabilistic risk analysis that, though its conclusions remain uncertain, is able to use data about real past events to calculate a probability. Surveying the expert community can tell us the probability that a given expert will believe in human extinction by AI, not the probability of the event itself. Given Bostrom's hedging, we are not just dealing with uncertainty but, what is more abstract, rhetorical play with *uncertainty about uncertainty*. We would have to be very generous readers to grant that there is some truth to these forecasts when the author is explicitly telling us not to read much into them. Given

15. Bostrom, 21.

the fundamentally speculative nature of extinction scenarios, such guesswork is understandable. The problem is when it is *also* treated as "scientific" and "rational" grounds for political policy—even as the only grounds worth mentioning.

Stepping Back from the Precipice

A similar problem is found in Toby Ord's use of ostensibly Bayesian probability in his recent book *The Precipice*. Ord's analysis draws on a form of probability theory that, in Ian Hacking's words, "offers a way to represent rational change in belief, in light of new evidence."[16] For Hacking, Bayesian theory matters most to "personal probability" or "belief-type probability." It is a way of formalizing the process of updating one's beliefs—or guesses about probability— by learning from experience. The basic idea is that given $Pr(H_1)$ as a "prior probability," new evidence E will give a posterior probability, $Pr(H_1/E)$. The process is iterative. The posterior probability can be fed back into the equation as a new value for H, and evaluated again in terms of new evidence, and so on. There are further complexities that we leave aside. The point is that new evidence is necessary for this looping process that offers a good way to incorporate prior beliefs—that is, to avoid the blunt tool of an empiricism that claims to think only with firsthand sensory evidence.

Like Bostrom, Ord is discussing the probability of AI-caused human extinction, but in the context of a general chapter about "quantifying risks" in the twenty-first century "risk landscape," so his statements about method apply to the field's basic model as well. For Ord, the greatest risk in the next hundred years comes from AI. He rates the risk of extinction at one in ten, then notes that such a high number for such a "speculative risk" needs more explanation:

A common approach to estimating the chance of an unprecedented event with earth-shaking consequences is to take a skeptical stance:

16. Hacking, *Introduction to Probability*, 171.

to start with an extremely small probability and only raise it from there when a large amount of hard evidence is presented. But I disagree. Instead, I think the right method is to start with a probability that reflects our overall impressions, then adjust in light of scientific evidence. When there is a lot of evidence, these approaches converge. But when there isn't, the starting point can matter.[17]

In an endnote to this passage, Ord goes on to explain that this approach to probability theory is Bayesian: it entails "starting with a prior and updating in light of the evidence," and the starting point matters.[18] Like Bostrom, he bases the updating calculation on "the overall view of the expert community that there is something like a one in two chance that AI agents capable of outperforming humans in almost every task will be developed in the coming century."[19] Unlike Bostrom, Ord does not include numbers from surveys or a citation, but he asks us to take his word about the AI expert's beliefs, then to accept his leap from the idea of human-level AI to AI as an extinction threat. While all of this is cast in the language of Bayesian updating, the Bayesian calculations themselves are promissory—as they must be given that we have no evidence to fill the variable "E" and temper our initial beliefs. So Ord's calculations are based on his own beliefs as an existential risk theorist and on the opinions of the uncited experts; Bostrom's calculations are based on his own beliefs and the average of a small sample of expert opinion. Arguably these probabilities are *all prior*. The authors give lip service to a Bayesianism that cannot be operationalized by definition for an unprecedented event.[20]

One could imagine a situation in which these ideas were more innocently speculative, so that the effort to treat guesswork as a

17. Ord, *Precipice*, 168.
18. Ord, 379.
19. Ord, 168.
20. In short, it's not Bayesian if you don't know if E is true. At best, if you want to be sympathetic, you could *assume* E is true and then reason like a Bayesian. Hacking's remarks on Bayesian probability and the examples he gives in *An Introduction to Probability and Inductive Logic* are in agreement with this perspective about "E."

foundation would be less problematic. But the mixture of utilitarianism and the rhetoric of probability that forms the field's basic model entails norms about what should be done in the present to mitigate existential risks. These promissory probabilities and their norms don't take place in a political-economic vacuum, isolated from practice. The Future of Humanity Institute, The Future of Life Institute, and the Centre for the Study of Existential Risk have received considerable funding, interest, and participation from tech billionaires. YouTube videos by Bostrom are widely viewed. As their institutional and media footprints suggest, this field is not at all shy about capturing attention and funding. After offering his belief-type probabilities, Ord argues that humanity has a 1/6 chance of extinction over the next hundred years, and lowering this number should be "a key global priority."[21] All five of the major existential risks—nuclear war, climate change, other environmental damage, AI, and designed pandemics—"warrant major global efforts on the grounds of their contribution to existential risk" (169). But the long menu of "minor" extinction risks merits major action as well. Given that the destruction of "humanity's entire potential" would be "so much worse than World War II," Ord suggests that it would be justifiable to create a global body for mitigating total existential risk by analogy with the UN. Another possibility is to "create a body modeled on the IPCC, but aimed at assessing existential risk as a whole."[22] For existential risk theorists, then, the policy and budgetary changes based on their rhetoric of probability ought to be dramatic and immediate. And when it comes to the popular nonacademic writings and declarations, the promissory calculations recede even further into the background, from appendices and box essays to gestures about the future of the field. Perhaps this is where the rhetoric of probability becomes most problematic. Given the public-facing nature of existential risk, this problem can hardly be seen as a nec-

21. Ord, *Precipice*, 169.
22. Ord, 396.

essary side effect of simplifying specialist work. Quite simply, the core practice of the field is not calculation but commentary about future calculation with normative weight in the present.

Radical political change is needed to ease the effects of global warming and preserve biodiversity, even if they are not extinction risks to humanity. But existential risk studies take such a bird's eye view of humanity that it is incapable of intervening in a context of austerity measures for education and public services, systemic racism, and extreme, worsening class inequality, among other social problems. So these genuine proposals for a massive expansion of existential risk mitigation must compete with other sorely needed fields that are at constant risk of being defunded. In the university space, existential risk studies gather steam and institutional support while austerity reigns for traditional humanities disciplines as well as for more recent fields like critical race studies that educate people about human histories of oppression in a time of nationalist and white-supremacist reaction. The coronavirus pandemic has highlighted the racial inequality of infection and death rates just as it has shown the need for radical change in our agro-industrial food systems. We need research at multiple scales across the sciences and humanities into these existentially crushing structures of oppression, how current relations between cultures and natures became disempowering for so many, and how both can change in the near future.

Existential risk analysis is not risk analysis. This statement is only slightly hyperbolic. To be fair, there are some cases where calculating rough probabilities of planetary extinction seems possible. The frequently cited example is asteroid strikes. But even Ord admits the theoretical limits inherent in calculating the probabilities of unprecedented events—especially the kind that, by definition, would leave no survivors to perform Bayesian updates. From Bostrom and Ćirković's early *Global Catastrophic Risk* to the growing archive of relevant papers on the website of the Centre for the Study of Existential Risk, our reading suggests that there is very little actionable probabilistic risk analysis going on in this field. Perhaps this is due to conceptual incoherence: the study of existential risk is founded on an impossible data point. Most of the extinction events that form its core object of study are simply not tractable in its model. We can study extinction through other methods, but we cannot estimate the probability of most extinction events—especially the kind that matter most to these theorists, which are the "human-caused" or technological ones over which "we" would seem to have the most agency.

The problem with the lion's share of this discourse is that it gains legitimacy by offering quantification in an academic and corporate research world that privileges quantitative fields. But what it really offers is typology, speculation akin to scenario planning, and—most important for this chapter—a promissory *rhetoric* of probability rather than rigorous calculations. Moreover, valuing the future of humanity quantitatively is not possible because such value is relative and qualitative.[23] But this has not in the past stopped attempts to evaluate humans quantitatively, and this evaluative "logic" appears incapable of extricating itself from scientific racism, biopolitical management, and eugenics. Bostrom himself casually examines the application of eugenics toward creating "greater-than-current-human intelligence" in *Superintelligence*.[24] He notes that "any attempt to initiate a classical large-scale eugenics program, however, would confront major political and moral hurdles" (36). But this is not a condemnation of such schemes, more an assessment of their inefficiencies and unlikeliness to work. Bostrom does point out that state-run breeding programs might also be used to produce docility and control, but nevertheless concludes his brief analysis of eugenics by noting, "Progress along the biological path is clearly feasible" (44). If the study and mitigation of existential risk were to become what its practitioners envision, then it is difficult to see how these blunt tools would avoid creating new ways of instru-

23. See, for example, the equations offered in Toby Ord's appendix "The Value of Protecting Humanity" in *The Precipice*.

24. Bostrom, *Superintelligence*, 36.

mentalizing lives in the service of a future they will never see with their own eyes. The tools of immunity have a tendency to attack what they are supposed to protect.

Extrapolation and the Unprecedented Event

In Bostrom's work, as we noted above, only the most extreme risks count as existential risks, the ones that constitute an extinction threat for humanity or for "Earth-originating intelligent life." Many of these risks share something in common with discourses about climate change and ecological collapse, as with nuclear war discourse before them: the fear that we are heading toward an apocalypse that will mean the collapse of "civilization" (a dated term that appears often in Bostrom's work and makes it reminiscent of grand narrative, "big picture" historians such as Arnold Toynbee, Oswald Spengler, Jared Diamond, and now Noah Yuval Harari). In order to "firm up where the boundary lies between realistic scenarios and pure science fiction," as Rees puts it, these scenarios must be possible and plausible even if their probability seems low.[25]

As we have seen, there may be no way to calculate the probability of most of these scenarios, with the exception of examples such as a terminal asteroid strike. Here the geological record and astronomical observation can provide evidence, and the possible asteroid event already has precedents like the K/T extinction event of about 66 million years ago. Indeed, the study of fossil records of past extinctions beginning in the late eighteenth century drove the first wave of naturalistic, secular apocalypticism. With such precedents, as we saw above, one has a chance of estimating the probability that a strike will happen during a given period of years. Without them, the estimate becomes a different kind of probability altogether. As Carl Sagan wrote about the threat of nuclear

25. Rees, foreword to Torres, *Morality, Foresight, and Human Flourishing*, 15.

war: "Theories that involve the end of the world are not amenable to experimental verification—or at least, not more than once."[26] To elaborate our critique above, when existential risk analysts do suggest numerical probabilities, what they do is offer a *probability* with a baked-in assumption of *possibility* or *plausibility*. There is no distinction between events that require such speculation and those that do not. There are even efforts to argue that such distinctions are untenable by merging them into a continuum.[27]

From the perspective of a critique of existential risk concerned with the epistemic structures that shape what it can say and its relations with science, there remains the question of what the field might be if not risk analysis. The question ultimately hinges on the way existential risk handles the logic of an unprecedented event that compels us retroactively, as though it has already happened. This logic is similar to what Ray Brassier calls "posteriority" in his own discussion of extinction: because the sun will burn out and the universe ultimately flatten in entropic dissolution, "the subject of philosophy must . . . recognize that he or she is already dead" and that "the posteriority of extinction indexes a physical anni- hilation which no amount of chronological tinkering can trans- form into a correlate 'for us,' because no matter how proximal or how distal the position allocated to it in space-time, it has *already* cancelled the sufficiency of the correlation" between subject and object.[28] We return to Brassier in the next chapter. What needs discussion here is the distinction between events about which we can extrapolate through what Hacking called "the taming of chance,"[29] and events that involve pure speculation like the arrival

26. Jill Lepore, "The Atomic Origins of Climate Science," *New Yorker*, January 30, 2017, https://www.newyorker.com/magazine/2017/01/30/the -atomic-origins-of-climate-science.

27. Bostrom and Ćirković, *Global Catastrophic Risks,* 6.

28. Ray Brassier, *Nihil Unbound: Enlightenment and Extinction* (New York: Palgrave Macmillan, 2007), 239, 229.

29. Hacking, *Taming of Chance.*

of genocidal aliens. By comparing extrapolation with posteriority, we get a fuller picture of the epistemology behind existential risk.

The anonymous AI researcher cited by Khatchadourian in the *New Yorker* conveys something more than glibness when they mention that they "don't worry" about AI ending the human species "for the same reason I don't worry about overpopulation on Mars."[30] This is another of the many sci-fi-adjacent thoughts that abound in the field (more below). The existential risk analyst might say that this remark just indicates a dangerous, cross-that-bridge-when-we-come-to-it attitude about something that could result in total human extinction. But another reading is more instructive when it comes to epistemic structures that should be more explicit in the work of existential risk analysts. The remark could also be about the difference between probability and possibility that so often slips into the blind spot of their work. The remark could mean that extrapolation from a place where we don't even know if human-level AI is possible to a place where it becomes superior and genocidal is an absurd counterfactual. One can think it, like thinking overpopulation on Mars. But how much should we worry about it or build our politics around a risk when we don't even know it to be possible in the first place? To do so would be to cross a threshold from extrapolation to another method entirely.

Thus a key question for any critique of existential risk is just what this "threshold" is and how it matters politically. After all, thinkers like Bostrom, Ord, and Torres might be right that although existential risks feel counterintuitively vague, impractical, and temporally distant, this does not necessarily make them unimportant—it just means that we have not evolved or learned the mechanisms to cope with them if not control them. From this perspective, long-term thinking seems salutary. Isn't turning "long-range forethought" into "a moral obligation for our species" what environmentalists have always wanted?[31] If so, where does realism grade into fantasy, and how well defined is the line? How should the thresholds between that which is tractable to probability theory and speculation about possible impossibles factor into nascent critical theories of extinction? To answer these questions, we need to continue down the path of positioning existential risk in its epistemic context.

One way to critique the "risk" models of existential risk might be to treat them as *scenario planning* rather than risk analysis. As Torres notes, existential risk studies can be traced back to the historical and technological conditions of the Cold War, which gave rise to a version of futurology geared to planning for possible futures without being able to predict them. The main risk and context was of course nuclear war, which is why existential risk authors often evoke Carl Sagan, who popularized the dire scenario of nuclear winter in the early 1980s. Going further back, another important precursor is Herman Kahn, the Cold War scenario planner and think-tank guru who worked for the RAND Corporation (a nonprofit funded in part by the U.S. Department of Defense). In recent years, scholars of the environmental humanities, anthropology, critical race theory, and speculative fiction have discussed this form of forecasting in some detail.[32] For anthropologist and Foucault scholar James D. Faubion, there is a "scenaristic" rationality that distinguishes this kind of governmentality from the statistical modes of risk society, which is fundamentally biopolitical. The scenario planning of figures like Kahn and Pierre Whack of Royal Dutch Shell is different and "parabiopolitical" because of its narrative method.[33] These two

31. Khatchadourian.

32. See, for example, Lynn Badia and Jeff Diamanti, eds., *Climate Realism: The Aesthetics of Weather, Climate, and Atmosphere* (forthcoming); Lindsay Thomas, "Forms of Duration: Preparedness, The *Mars* Trilogy, and the Management of Climate Change," *American Literature* 88, no. 1 (2016): 159–84; Kara Keeling, *Queer Times, Black Futures* (New York: NYU Press, 2019).

33. James D. Faubion, "On Para-biopolitical Reason," *Anthropological Theory* 19, no. 2 (2019): 219–37.

governmentalities, biopolitical and scenaristic, are not mutually exclusive. Given its tenuous grip on probability and its deep interest in apocalyptic narratives, perhaps existential risk is scenario planning under the quantitative guise of risk analysis.

R. John Williams makes points conducive to this interpretation when he discusses such "futurology" in fine-grained archival detail in "World Futures." His striking discussion of the scenario planning adventures of RAND and Shell during the 1950s and 1960s shows how they

> initiated a new mode of ostensibly secular prophecy in which the primary objective *was not to foresee the future but rather to schematize, in narrative form, a plurality of possible futures.* This new form of *projecting forward*—a mode I will refer to as World Futures—posited the capitalizable, systematic immediacy of multiple, plausible worlds, all of which had to be understood as equally potential and, at least from our current perspective, nonexclusive. It is a development visible, for example, in a distinct terminological transition toward futurological plurality.[34]

There is more to say about this approach and its contextualization in Cold War and corporate extractivist resource dominance agendas. For us, what it clarifies is a distinction between probabilistic risk analysis and a narrative method that schematizes possible futures, *embracing their plurality and unforeseeability.* The probabilistic understanding of future worlds does make existential risk similar to the kinds of corporate futurology discussed by Williams: for Bostrom and company, the point is that we cannot and should not ever know which of these futures will materialize as *the* extinction event; rather we should focus on lowering the probabilities of every imaginable extinction scenario that lurks in the future of humanity. The IPCC of existential risk would always be monitoring and updating (with what evidence is often unclear) the probabilities. What Eva Horn writes about "today's awareness of the future as

---

34. R. John Williams, "World Futures," *Critical Inquiry* 42, no. 3 (Spring 2016): 473; emphasis added.

catastrophe" applies here as well: "We are dealing with a metacrisis composed of many interrelated factors, dispersed into a multitude of scenarios, and distributed among many different subsystems."[35] Horn adds that these "metacrises" serve as staging platforms to make the future more legible and amenable to securitization and play out new norms of acceptable risk and reward. Even if existential risk does not play the same role in political economy as scenario planning, the former can most certainly attract funding and the attention of an audience savvy about the promissory, hedged, venture-capitalized futures of the start-up world. Another similarity is what Williams describes as the "ostensibly secular" and "quasi-theological" nature of World Futures. This strain of "ontotheology," to use Martin Heidegger's term, has clear connections to the apocalypticism of Bostrom and company, as with transhumanism's desire for immortality through the technological transcendence of finitude and embodiment.

But this is where the similarities end. The first difference between scenario planning and existential risk is that the latter's extinction scenarios are not to be understood as "equally potential," but rather differentially probable, with the differences intended to guide policymakers. A second difference is simply that scenario planning stays focused on the more limited and practical domain of "possible futures" and "plausible worlds."[36] Scenario planning seems just as tame as one would expect from the officious "masters of the world," those who aspire to present to an audience in the Pentagon or a skyscraper owned by Shell. Whereas existential risk, with its discussions of the simulation hypothesis side by side with doomsday prepping, hews to audiences familiar with escapist longings, life-extension research, and sci-fi fandom. Surely existential risk analysts would say that their approach is resolutely secular,

35. Eva Horn, *The Future as Catastrophe: Imagining Disaster in the Modern Age,* trans. Valentine Pakis (New York: Columbia University Press, 2018), 8.

36. Williams, "World Futures," 473.

and that the extinctions they discuss are entirely "plausible" and "possible." But as we've seen, these words mean something very different in this context than when they point to the near-future examples discussed by Williams.

The final difference to note is that existential risk does not imagine how we would act during its scenarios, since there would be no "we" to act. With scenario planning, even for nuclear war and obviously for energy-geopolitical economies of the future, the focus has been on developing plans of action that planners could imagine unfolding within each future world. They would hedge by anticipating multiple futures at once, futures that would not be mutually exclusive but might interlace and recombine in the complexity of a reality that always exceeds anticipation. But with existential risk, there is no acting within the scenario itself. Ord is quite explicit about this preemptive logic. He notes that *"necessarily unprecedented"* risks are especially difficult to work with because "by the time we have a precedent, it is too late—we've lost our future."[37] This means, again, that "to safeguard our potential, we are forced to formulate our plans and enact our policies in a world that has never witnessed the events we strive to avoid" (195). Existential risk analysts are in a difficult place: unable "to fail even once." To act on their probabilistic forecasts, they will have to take proactive measures: "sometimes long in advance, sometimes with large costs, sometimes when it is still unclear whether the risk is real or whether the measures will address it" (196). Existential risk analysis preempts; by definition it cannot guide our actions if the scenarios manifest. So the parallel breaks down on multiple fronts, and existential risk only partially overlaps with the paradigm of scenario planning or "World Futures."

There is something that needs further attention as we continue to understand the epistemic structures behind existential risk: the implications of Ord's concept of the "necessarily unprecedented."

37. Ord, *Precipice*, 195, emphasis added.

For existential risk theory to be as rational as it claims to be, it must depend on a kind of future anterior logic. We have to see our world as one in which the event of extinction has "already happened" or "always already might happen." The scale of its risk means that we should treat it as retroactively real. In turn, this realism of the present invoking a future gazing back on the past (a retroactive or posterior realism) makes the choice to act on low-probability risks seem rational. If we consider the foregoing arguments, it seems that existential risk hovers between two interlocking scientific realisms: (1) the realism of plausible, extrapolatable futures, and (2) the retroactive realism of an event so serious that, though "it is still unclear whether the risk is real," compels us to act *as if it were real* in the first sense of plausible extrapolation. This makes it something like a Kantian regulative idea.

Here we can see the difference from Brassier's concept of transcendental posteriority, where *inevitable* extinction retroactively shows the truth of scientific realism and nihilism. For existential risk, realism applies to what is possible but not yet actual. We are not "already dead"; we should act as though the total probability of extinction is high in order to prevent it. But this approach does correspond with Brassier's idea that "extinction is real yet not empirical": the consequences are great enough that we are asked to bracket skepticism about the distinction between probability and possibility and do without empirical observation.[38]

Wagers and Thresholds

Such scientific realism does not, however, mean that deep histories of religion are easy to escape, as we saw too in Williams's comments about the "quasi-theological" aspect of World Futures. Indeed, Ord's concept of the necessarily unprecedented event effectively rein-

---

38. Brassier, *Nihil Unbound*, 238.

vents Pascal's wager for a secular eschatology.[39] Pascal's wager is a troubling argument in favor of faith in God. For Pascal, it is rational to believe that God exists, or at least to try even though we can't know one way or another, because if he doesn't the consequences will be finite: the loss of some pleasures, some freedoms, and so on. But if God does exist, then the rewards for belief are infinite, as are the punishments for disbelief. So the rational choice is to try to have faith and bet on the infinite rather than the finite. The rationality of the choice *will have been verified* from the perspective of a future reality if one dies and learns that God existed all along. Otherwise, one just dies. In the meantime, all we have to go on is the difference of kind between outcomes: finite and infinite.[40]

Despite being a secular eschatology based on probability rather than a theological one based on metaphysics, existential risk shares something in common with Pascal's wager because it suggests that risks with extremely low probabilities should be treated as clear and present dangers. Indeed, Bostrom cites the wager in *Superintelligence,* and he has published about it elsewhere, going as far as to transform Pascal into a devotee of a utilitarianism that seeks to maximize "astronomical" or "infinite" value.[41] Another use of this cosmic wager logic appears in Sir Martin Rees's foreword to Torres's book *Morality, Foresight, and Human Flourishing,* where Rees writes, "the stakes are so high that those who are involved in [the study of existential risk] *will have earned their keep* even if they reduce the probability of catastrophe by a tiny fraction."[42] The risk is calamitous enough that it bears an analogy with Pascal's distinction between finite and infinite (and here we recall Derek Parfit's argument that losing 99 percent of humanity would be catastrophic but losing 100

39. Thanks to historian of math Michael P. Barany for this suggestion.

40. Blaise Pascal, *Pensées,* trans. W. F. Trotter (London: Dent, 1910), section 23.

41. Nick Bostrom, "Pascal's Mugging," *Analysis* 69, no. 3 (2009): 443–45.

42. Sir Martin Rees, foreword to Torres, *Morality, Foresight, and Human Flourishing,* 15.

percent would be infinitely worse[43]). Like Pascal's wager, Rees's statement involves a retroactive logic signaled grammatically by his use of the future perfect tense: existential risk analysts "will have earned their keep." Even if we do not know if a given extinction is possible, much less its probability, acting in a way that proleptically prevents an extinction event is the right thing to do, and society's resources will have been earned if the dice roll goes our way. Like the choice between infinite punishment and the same death that would have happened anyway whether you believe in God or not, it is rational to pay their salaries because these risk analysts are able to choose paths that lead away from Parfit's "infinitely worse" 100 percent death rate. So long as we're still here, they *might* have been successful. But if they fail, as Ord reminds above, by definition no one will be left to know it.

This is indeed a strange legitimation, a strange logic of verification, and a strange relation between the rational and the real. We are tempted to see it as the kind of overreach of rationalism common in analytic philosophy, where statements that can be simplified into logically coherent and noncontradictory forms are seen to have a privileged access to reality rather than just (as we think) illuminating a very decontextualized region of the space of reasons. But theorists of existential risk do make a strong case for the counterintuitive nature of their claims, suggesting that common sense militates against long-term thinking and "safeguarding" human existence. As we have tried to suggest throughout, some of their ways of stretching common conceptual frameworks and political norms to meet the cosmic scale of their topic are worthy of careful critique—not dismissal, but the effort to occupy and unpack the shadow cast by existential risk. For now, we elaborate how the logic of the unprecedented and retroactive rationality hinge on a *threshold* between the plausible and the thinkable.

---

43. Derek Parfit, *Reasons and Persons* (New York: Oxford University Press, 1984), 453.

The language of thresholds in complex systems has become common in discussions of climate change and Earth system science, and a number of scientists have recently argued, in high-profile publications, for the existence of planetary "tipping points" that might be an existential threat to humanity. If the climate were to pass such a threshold, it would stop warming in a linear way that reflects the rate of accumulation of carbon in the atmosphere. Instead, it would change in a sudden and nonlinear way, driven by self-reinforcing feedback loops that quickly spin out of control. Such "hothouse earth" scenarios create the greatest anxiety for those who think global warming might lead to the extinction of humans or even life on Earth, not just suffering or the collapse of modern societies.[44] Not all existential risks would involve the same kind of thresholds as radical climate change, but the concept is helpful because it illuminates an epistemic break between futures that can be extrapolated from the present and futures that are more speculative.

Alicia Juarerro makes clear how this notion of the complex threshold interacts with epistemic frames in an essay that draws on the work of Stuart Kauffman, where she writes, "It is impossible to predict emergent properties *even in principle* because the categories necessary to frame them *do not exist until after the fact*."[45] With this claim, which casts doubt on a strict division between ontology and epistemology because frames of knowledge depend on real emergences in complex systems, much depends on what Juarrero means by "prediction." For the purposes of the critique of existential risk, we note that her point goes beyond the idea that these emergences cannot be predicted *with certainty,* a limitation which is absolutely

44. See Will Steffen et al., "Trajectories of the Earth System in the Anthropocene," *Proceedings of the North American Academy of Sciences* 115, no. 33 (2018): 8252–59.

45. Alicia Juarrero, "What Does the Closure of Context-Sensitive Constraints Mean for Determinism, Autonomy, Self-Determination, and Agency?" *Progress in Biophysics and Molecular Biology* 119, no. 3 (2015): 510; emphasis added. See also Stuart Kauffman, *Humanity in a Creative Universe* (New York: Oxford University Press, 2016).

accounted for in the way existential risk's basic model processes uncertainty. She goes beyond this kind of probabilistically tractable uncertainty by introducing a break in the continuum that Bostrom and Ćirković argue for in *Global Catastrophic Risks,* precisely in the moment when they discuss the field's relation with science:

> Although more rigorous methods are to be preferred whenever they are available and applicable, it would be misplaced scientism to confine attention to those risks that are amenable to hard approaches. Such a strategy would lead to many risks being ignored, including many of the largest risks confronting humanity. It would also create a false dichotomy between two types of risks—the "scientific" ones and the "speculative" ones—where, in reality, *there is a continuum of analytic tractability.*[46]

The break that this continuum papers over is a crucial one because it suggests a limit on using the tools of probability to extrapolate from current conditions. Even probabilistic prediction is prediction. But the inability to predict certain emergent properties from initial conditions would suggest that the very parameters that would enable this kind of prediction, which are also crucial for climate modeling, are set to change. If Juarrero is right that the "categories necessary to frame" such emergent events do not exist until after the fact, then existential risk theorists are making category mistakes when they argue for extrapolative and retroactive realism.[47] If such thresholds exist and can be used to interpret the unprecedented events of existential risk analysis, then there is no way grope toward them from the present nor use them retroactively (in the mode of Pascal's wager) to compel belief and action today. The very concept of an unprecedented event demands that we conceptualize a break in the continuum that stretches from past to future, known to unknown.

We think it is important to emphasize the breaks in the continuum between "scientific" and "speculative" risks. There may be

---

46. Bostrom and Ćirković, *Global Catastrophic Risks,* 6.
47. Juarrero, "What Does the Closure of Context-Sensitive Constraints Mean?," 510.

others, but we focus here on one kind of break in order to flesh out the logic of the unprecedented event. Here our approach is "critical" in the sense that it reverses the process of collapsing differences into an identity (a continuum suggests oneness) that then serves as a standard for rationality and political action. If existential "risk" continues to be well-funded and impactful outside the academy—and if it takes on even a fraction of the power its adherents think it should—then critique of the field's reductions will be all the more important.

Science, Scientism, and Science Fiction

To think through the distinction between scientific and speculative is to return once again to problems of demarcating what counts as science and what does not in a time when the social authority of the sciences is both greater than that of the humanities and treated with skepticism (as the pandemic has shown) by a growing list of "post-truth" political projects. When Bostrom and Ćirković argue against "scientism," the idea that a monolithic Science is the only kind of knowledge that matters, they are marking out a "rational" space for risk analysis that cannot be limited to things about which we can collect evidence. They rightly eschew scientism. But their mathematical metaphor of the continuum proceeds to bring science back into proximity with their approach. If scientific and speculative risks are on a continuum, then Bostrom and Ćirković are implying that their means of reasoning about extinction might become scientific after all.

By claiming science and reason for their own approach to extinction, such arguments also make it more difficult to open what counts as knowledge to a multiplicity of ways of knowing beyond reductive notions of "science" and "reason." One way is the kind of existential ecology that we discuss in more detail in chapter 3. Another is science fiction, which we might have given more space in this book. Such ways of knowing the world seem much better suited to topics like future extinction and cosmological scales than the mixture of probability and utilitarianism at work in existential risk, but they lack the legitimacy that attaches to the quantitative. Scientism might well be blocking our intellectual cultures from addressing some structures of existence that have the potential to guide alternatives to cynical individualism and toxic short-term thinking.

In the world of debates about literary genres associated with science, one relatively recent distinction is surprisingly similar to one we tried to make by insisting on a break in Bostrom's continuum. Here, too, the logic hinges on where extrapolation to plausible futures gives way to more remote speculation. Margaret Atwood distinguishes "speculative fiction," the name for novels that deal with the near future in terms of socially relevant extrapolation, from "science fiction," which entertains fantastic powers beyond the ken of known science such as time travel or magic monsters. Hannes Bergthaller finds that the distinction between speculative fiction and science fiction "hinges on the realism and probability of the fictional world depicted in a given text: while the latter supposedly deals in 'things that could not possibly happen' . . . the former is concerned with 'things that could happen but just hadn't completely happened when the authors wrote the books.'"[48]

One influence on the recent burst of interest in climate fiction follows a clash between Ursula K. Le Guin and Atwood that took place in the 2000s. Atwood marks a difference between the kind of science fiction she considers politically useful and the more fantastical regions of the genre. She distinguishes between the impossibilities of "science fiction proper" and "speculative fiction, which employs the means already more or less at hand, and takes place on Planet Earth."[49] But in a 2009 review of Atwood's *The Year of the Flood*, Le Guin takes issue with Atwood's distinction. Le Guin makes her

48. Hannes Bergthaller, "Cli-Fi and Petrofiction: Questioning Genre in the Anthropocene," *Amerikastudien* 62, no. 1 (2017): 123.

49. Margaret Atwood, "*The Handmaid's Tale* and *Oryx and Crake* in 'In Context,'" *PMLA*, 119, no. 3 (2004): 513.

point comically, writing that she *would* like to review Atwood's novel using "the vocabulary of modern science-fiction criticism, giving it the praise it deserves as a work of unusual cautionary imagination and satirical invention."[50] But since Atwood insists that she is not writing science fiction, Le Guin constrains herself to Atwood's "wish," using only "the vocabulary and expectations suitable to a realist novel." What follows is a dry evaluation of the novel based on its plausibility. For example, the flood itself (a pandemic) is an "abstraction, novelistically weightless," and the characters lack three-dimensional, Austenean depth. Le Guin's target is the high-culture pretense of distinguishing speculative fiction from science fiction, devaluing science fiction like her own, which lacks realism, as less serious, less political (Atwood also calls speculative fiction "social science fiction"), lacking aesthetic quality, and thus relegating it to "the genre still shunned by prize awarders . . . the literary ghetto."

What the clash between Atwood and Le Guin illustrates is an effort, on different terrain, to process the threshold between plausible extrapolation and speculation about unprecedented events. If such a recent, formal, and literary distinction is relevant to our study of existential risk's relation to science, then what is the role of science fiction in the critique of existential risk? Some points about science fiction help to set up a concluding attempt to deconstruct what Ord calls the "basic model" of existential risk.

The textual ties between existential risk and science fiction are peppered throughout the literature we've been citing. Science fiction, in turn, has always been in dialogue with human extinction, AI, and other scientific speculation about the future. One reviewer of Bostrom and Ćirković's *Global Catastrophic Risks* encapsulates the book as "risk assessment meets science fiction." As Rees puts it, "There needs to be a much expanded research program, including natural and social scientists, to compile a complete register of

50. Ursula K. Le Guin, "*The Year of the Flood* by Margaret Atwood," *Guardian,* August 29, 2009, https://www.theguardian.com/books/2009/aug/29/margaret-atwood-year-of-flood.

possible 'x-risks,' to firm up where the boundary lies between real-istic scenarios and pure science fiction, and to enhance resilience against the more credible ones."[51] The literature of existential risk is sprinkled with both disavowals of science fiction (in the form of "*x* scenario is not just science fiction, but something to take seriously") and conveniently picked references to science fiction authors, but it utterly ignores sci-fi scholarship. Such an absence of engagement could be the result of long-standing conflicts between continental theory and analytic philosophy, imagination and reason; it could reflect the low profile of literary studies from the perspective of humanities fields that hope to receive the blessing of the sciences, or even to be included, themselves, in the scientific pantheon. In any case, there is in existential risk a tendency to use the word "rational" in the fully positive sense and "irrational" in the fully negative, without the kind of questioning that comes from prag-matism, phenomenology, the Frankfurt School, postcolonialism, feminist theory, or critical race studies.

Disciplinary quibbles aside, however, the most pressing issue for our critique is that this omission of the entire preexisting lit-erature on science fiction is, for existential risk, symptomatic of an untheorized relationship to science fiction as a genre. Science fiction is "fetishistically disavowed" throughout the literature of this field that offers itself up as a science of the unobservable, or at least a rational approach to it. In Freud's sense, fetishistic disavowal means repeatedly casting something outside the sphere of one's argument or way of life in a way that suggests that one is actually obsessed with it or desires it on an unconscious level. "Bostrom dislikes science fiction."[52] In our case, the concept does not apply to the psychic lives of the risk analysts but to the discourse of ex-istential risk more broadly. It seems that science fiction cannot be taken seriously, but that it is always in the blind spot of this field.

51. Rees, foreword to Torres, *Morality, Foresight, and Human Flourishing*, 15.
52. Khatchadourian, "Doomsday Invention."

Perhaps what is really being disavowed here, at a theoretical level, is not any discipline or genre, but rather the shift in theory and practice that should take place, but doesn't, when existential risk theorists speculate about what they can't possibly know.

We look at one such shift in theory and practice in the next chapter: from existential risk to existential ecology as an approach to extinction. Another shift would be a thorough focus, already well established, on narratives of apocalypse and utopia and on the work of scholars who have been studying these modes for decades. Horn's *The Future as Catastrophe* is particularly attentive to the fact that "knowing and communicating about the future is impossible without stories: stories that 'look back' from the future to the present or that extrapolate from past predictions about what is to come."[53] She argues that these narratives, especially science fiction and scenario planning, "structure the way we anticipate and plan for the future and, above all, how we try to prevent catastrophic futures from occurring" (10). She pushes this claim further when she writes that "fictional scenarios of the future in literature, film, popular culture, and popular nonfiction . . . are neither mere symptoms of the collective psyche nor simply media of ideological indoctrination but *epistemic tools* to understand and discuss potential futures" (10). In so doing, they reveal something that already exists in the present while priming the imagination so that drastic social changes can become conceivable. Horn is in agreement with Ursula Heise's point that "the basic strategy of science fiction is to present our own society as the past of a future yet to come."[54] This is only a glimpse of what science fiction scholarship does with futures of extinction and utopia, but it provides an alternative to the idea that science and reason are the only forms of knowledge that count.

Research like Horn's takes a qualitative and narrative approach to the same topic that preoccupies existential risk theorists. In this

---

53. Horn, *Future as Catastrophe*, 10.

54. Ursula K. Heise, "Science Fiction and the Time Scales of the Anthropocene," *English Literary History* 86, no. 2 (Summer 2019): 282.

moment of quantitative dominance, when statistical methods are applied even to topics that seem utterly intractable to them (when "predicting the unpredictable and empirically studying the unverifiable" seems viable to some[55]), qualitative (aesthetic and critical) studies might seem unrigorous to both scientists and policymakers. But they should be taken seriously, especially when the object of study is so constitutively fictional. This would mean embracing the idea that fiction is a source of knowledge, not just a convenient reference.

In light of this chapter's critique based on existential risk's relations with science, we can sum up and draw some conclusions that question its modeling practices and presuppositions. The basic model treats extinction probabilistically, applying risk analysis concepts used by insurance companies and war games, among other industries, then combines it with a utilitarian approach to ethics. Any risk can be studied in terms of its probability and its potential damage to humanity, often calculated in terms of humanity's entire future potential. By adding this deep-time calculus of the value of humanity to the question of how we should make political progress today, the basic model opens itself to assumptions about what humanity will be like far into the future and uses these as the basis for how we should act now. We argued that the field's use of probability is more often rhetoric of probability that promises future calculation, and we questioned the value of quantitative politics for the study of extinction. By looking at the extrapolative and futural scientific realisms of existential risk, we argued that it ignores the thresholds across which extrapolation of possible worlds from the conditions of the present must fail and turn into speculation. Yet existential

55. Thomas Rowe and Simon Beard, "Probabilities, Methodologies and the Evidence Base in Existential Risk Assessments," working paper, Centre for the Study of Existential Risk (2018): 2, https://philpapers.org/rec /ROWPMA-6.

risk also makes its claim to realism retroactively, looking back from a necessarily unprecedented future event to argue about what is real and what we should rationally act on today. In scenario planning and science fiction, we saw two closely related genres, wired in parallel, as it were, with existential risk. These genres show alternate methods that often cryptically and obliquely shape what is meant to be a rational and scientific modeling. In short, there are fundamental epistemological problems with the field of existential risk.

Yet studies of existential risk do evince an intriguing encyclopedic aesthetic in their need to cover every imaginable extinction scenario, whether or not we know it is possible. Behind the accessible, action-oriented writing of its proponents, there lurk weirder regions of philosophy, especially the kind that adjoins to logic and math. Alternative-world rationalists from Leibniz to Charles Sanders Peirce to David Lewis are only a few steps away from the field's publications. If they had gotten closer, of course, they might have pushed the policy and philanthropy professionals in a different direction.

There is also a current of scientific modernity that existential risk understands very well: the impact of chance and probability, whether we see it through the lens of David Hume, the rise of statistics, risk, and biopolitics, the rise of statistical mechanics in physics, or the role of probability in quantum theory. We can imagine an existential risk chapter in Dennis Danielson's *The Book of the Cosmos,* which chronicles western cosmologies from the ancient Greeks to Einstein and Hawking to the "multiverse."[56] The cosmology of existential risk would have very little interest in origins but remains fascinated with churning out every possible end for Earth, life, or humans. The adherents of this cosmology would constantly cycle from one scenario of the end of the world to another, as though they were numbered balls drawn from a Bingo

56. See Dennis Danielson, *The Book of the Cosmos: Imagining the Universe from Heraclitus to Hawking* (New York: Basic Books, 2000).

machine. Each conflagration would appear more or less often at a rate set by its constantly updated probability. The story could go on, much like in Olaf Stapledon's *First and Last Men* (1930), which narrates multiple civilizational beginnings and endings over a two billion-year span on Earth. The intriguing thing about all of this is how existential risk generates counterintuitive ways of linking the past and the future on the terrain of probabilistic ontology.

The idea of an institution dedicated to the deep future or to possibilities that might not be possible sounds wonderfully decadent and intriguing. One could imagine the novels and films that would address it by combining plots about technocracy, mixed with Borgesian thought experiments to account for possible worlds. This institution would be a kind of grand, metascientific, interdisciplinary priesthood of extinction (remember Ord's humble analogy with the IPCC), auguring the improbable events and courses of action they might prevent through mathematically and rationally intricate manipulations of current knowledge, which surely resemble some financial tools used to hedge bets on the market. Bostrom seems to traffic in a new kind of grand narrative invested in the idea of predicting and determining the course of history. But it is less a grand narrative march toward transhumanism than a vast, fragmentary array of (im)possible worlds. The scope is as big as it can be, but the logic is counterintuitive and open-ended, not totalizing. There are ways to repurpose some of these ideas for an approach to extinction that would be more critical about its existential structures. Perhaps a less instrumentalized approach to the limits of the modern cosmic worldview would help in reinventing the discouraging (especially as we write, in 2020) project of achieving social and environmental justice.

3. The Existential Roots of Existential Risk

STRANGELY, not in any interview or essay by anyone in the field of existential risk have we seen a discussion of what "existential" means. For these philosophers, the term existential is treated in a simplistic way as a synonym for the totality of human existence put into crisis. None of the contributors to the field of existential risk consider how the term is caught up with the legacy of existential philosophies, an avoidance that turns into more than a missed opportunity. This dismissal is not explained away by noting long-standing rifts between continental and analytic philosophy. On the contrary, we would expect such an explicitly interdisciplinary project as the study of existential risk to discuss variations of its central concept. Bostrom's subject of existential risk, at first, appears drawn from "classical" existentialism in that this individual is deeply anxious and preoccupied with mortality—with examining all facets of dying or how to process the subject's "being-toward-death." But for Bostrom, mortality itself is among the problems of existential risk that need to be resolved by pursuing a posthuman existence. His version of the posthuman apparently enjoys all that the existential subject has to offer but without being burdened by the existential condition itself.

Here we offer a sketch of what existential thought entails in a broad sense, without narrowing the field to one philosopher or school. And from the outset, existential thought need not be treated as exclusive to the human species.[1] We include a number of criticisms of existential thought as revisions and new developments rather than rejections of the need to tarry with how embodied "lived experience" resonates across the philosophical field. The problems with existential thought are many, but the premise of elaborating a robust philosophy of precarious embodiment still matters. In its most basic aims, existential philosophy examines the relationship of lived existence to the broader conditions of being, posing this relation as central to philosophical inquiry. Such thought starts from embodied existence rather than abstract presuppositions of God or Nature, or proof from empirical data, a priori reason, or metaphysics. Existential embodiment is not grounded in naturalism or any fixed historical purpose (and the same goes for existential oppression). To be existentially embodied means that we come to know ourselves and the world through relations and connections (experiential and conceptual) that are continually negotiated and not guaranteed by preset metaphysical principles. Existential embodiment is thus inherently "risky." This philosophical method throws immediate light on how some bodies more than others—bodies that are marked as racially or socially inferior—are treated as more risky or caught in cycles of disempowerment. Mitigating existential risk makes sense not in the abstract but only in confronting concrete situations of suffering evident in instances such as colonialism, racism, sexism, economic oppression, and the domination of peoples and environments.

Since there are many summaries of existential thought available, our purpose here is to provide a brief analysis of existential think-

---

1. See Tom Regan's notion of animals as "subjects-of-a-life" in *The Case for Animal Rights* (Berkeley: University of California Press, 2004), Cary Wolfe's account of animal *Dasein* and biopolitics in *Before the Law: Humans and Other Animals in a Biopolitical Frame* (Chicago: University of Chicago Press, 2012), and Michael Marder's theory of plant phenomenology in *Plant-Thinking: A Philosophy of Vegetal Life* (New York: Columbia University Press, 2013).

ing in connection to existential risk. Histories of philosophy credit Schelling for introducing the term "existence" (borrowed from the medieval concept of *existentia*)[2] in the modern sense as identifying the position of the subject contra the abstraction of the concept, a position further developed by Kierkegaard's philosophy of existence as "the subjective problem."[3] While the methods for this inquiry into subjective being varies among philosophers, philosophers of existence agree that the ways of knowing subjective experience can never be fully reducible to objective ways of knowing. First-person experience cannot be fully explained or imitated from the position of a neutral third-person observer (though this does not mean that subjectivity is entirely impervious to objective knowledge, just that there is no final, determinative science of the subject). The difference between subject and object need not be due to some unexplainable metaphysical property of dualism. Indeed, existential thought need not take a definitive position on the question of the "hard problem" of the mind as physical or metaphysical (this also is evidence of the limits of this philosophy—it is not meant to explain all of reality but to give a robust account of lived experience and interrelational being). Even those who adhere to physical/materialist principles do not just treat themselves and other humans as things. People recognize other humans as mortal persons with unique "lived experiences" whose status is different from objects, even if we grant this status may be just due to mechanistic complexity, or ultimately taken to be merely fictional or pragmatic.

The distinct contribution of existential thought to the space of philosophy is to insist that, for the subject, both subjective and objective perspectives are simultaneously possible and intertwined

2. Hannah Arendt, "What Is Existential Philosophy?," in *Essays in Understanding: 1930–1954,* ed. Jerome Kohn (New York: Harcourt Brace, 1994), 167.

3. Søren Kierkegaard, *Concluding Unscientific Postscript,* trans. David F. Swenson and Walter Lowrie (Princeton, N.J.: Princeton University Press, 1968), 115.

yet nonidentical to each other. This thought is logically paradoxical but true to what Simone de Beauvoir called the "irreducible ambiguity" of lived experience.[4] Existential thought does not reduce existence to the biological (the "given") or the cultural (the "made") but includes both. For example, in existential thought birth and death are not adopted as strictly naturalist categories but rather are converted into the existential categories of natality and mortality (or "finitude") that are understood not as reductive determinations but "conditions" of existence that require interpretation and engagement. It is not that the subject as animal-human-machine is natural insofar as it is animal—aspects of nature and culture are intertwined in each of these categorizations of the subject. Each of these categories combines interpretive as well as determinative ways of knowing. Existential thought thinks the conditions of existence across this "cybernetic triad" as contradictorily coherent, both personal and impersonal, first person and third person. But existential thought is not merely subjectivism; each subject is unique as well as socially mediated and reliant on objective material reality. Subjectivity is experienced at the level of the body, in intersubjective relations with other subjects, and interobjective relations with the surrounding environs. Subjectivity is manifest at the level of self-determination and at the level of interconnectedness with other living and nonliving entities, a mix of determinacy and indeterminacy, individuality and sociality, spontaneity and automation, calculability and incalculability acting in simultaneous cooperation and conflict. There is no clear divide between these paradoxical overlapping yet distinct conditions—each side of the paradox is implicated in the other in what is called a double bind.

In existential thought, the life or death of the thinker plays a central role in what constitutes the philosophical field of the subject. Life and death define the core ontological structure of lived expe-

4. Simone de Beauvoir, *The Ethics of Ambiguity,* trans. Bernard Frechtman (New York: Philosophical Library, 2015), 8.

rience and serve as coordinates for a wide range of epistemological inquiries and phenomenological intentions. Furthermore, existential thought stresses the existentiality of its own terms. That is to say, philosophical concepts have existential conditions of finitude of their own in which their ongoing ability to exist and generate meaning or cognition is at stake. Language itself has existential properties—there are no permanent signs or perfect perpetuations of sense and communication. Existential thought assumes that there are no guarantees, no permanence or promised presence for any beings. The same precariousness applies to much of the space of philosophy itself. Contra logical positivism, existentialism took philosophy in the direction of becoming open to the undecidable and precarious relationship between referentiality and reality.

Existential thinking seeks to cognize the paradox of the concept as both universal and perishable. This thought does not deny the role played by the traditional architectonics of philosophical truth ascribed to a permanent and indestructible realm of thought, but it adds that the perishability, finitude, and precarity of existences and concepts are also true accounts of being in the world. As Adorno puts it, "Eternity appears, not as such, but diffracted through the most perishable."[5] In existential thought, there is nothing that offers metaphysical security for any lived existence or any concept in which meaningfulness is at stake, because meaning is to be generated precisely in the precarious, risky, exposed condition of a lived existence that has to actively make sense of the lack of full determination of its own finite life. Existence has a built-in ethics that requires it to care for itself as itself, since existence is not beholden to any biological destiny, metaphysical essence, or absolute historical task. Rather than trying to "solve" or explain away the subject and philosophy's precariousness (for example, by declaring it all just mechanistic activity), existentialism insists on tarrying

5. Theodor W. Adorno, *Negative Dialectics,* trans. E. B. Ashton (New York: Continuum, 1995), 360.

with the precarious as the very project of philosophy and embodied existence. The finitude of bodies and thoughts has the effect of expanding the space of existence and the space of philosophical reasoning—responding to Hegel's phrase that the philosophical system entails not just substance but also subject.

Existentialism and Extinction

Many of the first philosophers who foregrounded the existential subject as concerned with its own finitude and fragility understood how philosophy needed to reckon with the scientific evidence of extinction that implicated all life as precarious. Kierkegaard opens *Fear and Trembling* (1843) with the vision of human generations passing on like other natural cycles: "If one generation emerged after another like the leafage in the forest, if the one generation replaced the other like the song of birds in the forest, if the human race passed through the world as the ship goes through the sea, like the wind through the desert, a thoughtless and fruitless activity, if an eternal oblivion were always lurking hungrily for its prey and there was no power strong enough to wrest it from its maw—how empty then and comfortless life would be!"[6] Kierkegaard's response to the naturalistic view of extinction was to double down on the tenuousness of the human as philosophically primary. He proposed that the very singularity of the subject, having only faith and finitude rather than logical certitude and infinite assurance, was actually higher in standing than universal reasoning. The individual could suspend reliance on the rules of an already-assured universal ethics in pursuit of an even higher ethical devotion precisely because such subjective commitment could not relinquish responsibility by saying objective forces compelled behavior. Kierkegaard thought ultimately only faith afforded by Christian

6. Søren Kierkegaard, *Fear and Trembling and The Sickness unto Death*, trans. Walter Lowrie (Princeton, N.J.: Princeton University Press, 1968), 30.

theology, and not universal logic or contingent biological existence, could rescue the subject from existential oblivion.

Nietzsche, instead of retreating from the factual knowledge afforded by biology of the eventual oblivion of humanity, embraced a
"nonmoral" worldview in which humans played a bit part, opening
his essay "On Truth and Lies in a Nonmoral Sense" (drafted in 1873)
from a cosmic perspective that belittled human doings:

> Once upon a time, in some out of the way corner of that universe
> which is dispersed into numberless twinkling solar systems, there
> was a star upon which clever beasts invented knowing. That was the
> most arrogant and mendacious minute of "world history," but never-
> theless, it was only a minute. After nature had drawn a few breaths,
> the star cooled and congealed, and the clever beasts had to die.—One
> might invent such a fable, and yet he still would not have adequately
> illustrated how miserable, how shadowy and transient, how aimless
> and arbitrary the human intellect looks within nature. There were
> eternities during which it did not exist. And when it is all over with
> the human intellect, nothing will have happened. For this intellect
> has no additional mission which would lead it beyond human life.[7]

Nietzsche absorbed the astronomical insight of his time that the
sun would eventually die out, and with it, humanity (Ray Brassier
makes similar arguments below). From this cosmic perspective,
Nietzsche hypothesized the real as composed of chaotic forces
and eternal flux. Yet like Kierkegaard, Nietzsche affirmed subjective life experience as still worth pursuing, which for Nietzsche
meant advancing one's own will to power and self-creative "style."
Extinction and impermanence freed humans from needing to find
any higher meaning in suffering or self-limitation for the sake of
some abstract metaphysical principle (for Nietzsche this included
rejection of religion but also collective human projects like socialism
and democracy). In a world in which value and even "life" itself
had to be understood as a temporary and contested metaphor, the

7. Friedrich Nietzsche, "On Truth and Lies in a Nonmoral Sense,"
*Philosophy and Truth: Selections from Nietzsche's Notebooks of the Early
1870's*, trans. Daniel Breazeale (New Jersey: Humanities Press, 1990), 79.

only value left would be the heroic–tragic stance toward one's own subjective existentiality facing the chaos, while admitting no stable values could endure the flux.

Rejecting the biologism and subjectivism of Nietzsche, Heidegger contributed immensely to the elaboration of a philosophy of the subject that is neither subjectivist nor objectivist. This understanding of subjective existence would have a huge effect on existential thought, yet Heidegger's own political convictions toward fascism and anti-Semitism have prompted many readers of his work to use Heidegger's own ideas against his view of their implications. In *Being and Time* (1927), Heidegger goes through all sorts of concatenations with his baroque jargon to develop a theory of subjectivity that does not begin with the conventional subject/object or ego/world split. The figure of *Dasein,* Heidegger's term for the type of being that humans have, has a "pretheoretical" or nonconceptual attunement toward the surrounding world. Pretheoretical means that one is already embedded in a real-world situation that requires the subject to respond but is not created solipsistically by the subject or by a priori reason. *Dasein* does not first detect distinct objects disconnected from any context (a problem in the phenomenological method of Husserl) or axiomatic reasons abstracted from worldly conditions (as in logical positivism). Instead, *Dasein* already is oriented in the world in which there exists a horizon of concern and meaningfulness that provides the conditions for subjects, objects, reasons, or events to make sense. These conditions of existence are themselves precarious and in need of constant maintenance rather than transcendent or guaranteed to persist.

There are several reasons why this brief sketch of the existential condition detailed by Heidegger is of central importance to the current conceptualizations of existential risk. Heidegger charts the basic conditions of possibility for how a being is able to have a sense of care at all—a care that includes not only the concern for one's own possibilities and finitudes but also immersion in the world and the relationships that make up our shared condition. Heidegger finds care and finitude built into the ontological level, the very constitution of subjectivity: "As soon as Dasein expressed anything about itself, it has already interpreted itself as *care*."[8] Care is not just self-care, but always also implicated in care for being in the world and for being with others. Of course, people can be careless and neglect themselves and the world, but they cannot change the fact that their very ontology is imbricated with others.

Against the individualistic theories of the existential subject posed by previous philosophers (singular-mind or isolated ego-based philosophies including Descartes, Kant, Kierkegaard, and Nietzsche), Heidegger emphasized how existence entails being embedded in an intersubjective and interobjective world. Many subsequent philosophers, especially Lévinas, Merleau-Ponty, and Irigaray, argued that Heidegger did not pursue enough an analysis of the central role of intersubjectivity and its effects across all conditions of existence. What is most relevant here is that Heidegger's elaborations of existence provide a basis for how humans engage meaningfully with the world and with their own finitude. Care is grounded in the ontological relation of not just the subject to itself but also *Dasein* to the world. The *Dasein*-world relation is prior to subject/object dualism, while this simplistic dualism still plays a central role in utilitarianism that roots value in autonomous self-interested subjects. This subject/object dualism falters most when life straddles the gap and some existences and some life forms are treated dismissively as objects rather than agential subjects.

Heidegger refused to ground existence and ethics strictly in biology or in "objective" abstract reason and logical positivism, and he also rejected "naive" subjective-relativist positions. This philosophy of existence does not reduce the world to the individual ego's point of view, nor does it elevate subjectivity and consciousness to a cosmic scale, but begins with *bodies that care* as immersed within "pretheoretical" material–symbolic interrelations. Prior to treating

---

8. Martin Heidegger, *Being and Time,* trans. Joan Stambaugh (New York: SUNY Press, 1996), 171.

the objects and environs around us as mere things subject to physical as well as epistemic mastery, this argument finds "being-with"
and "being-there" as formative for relations of care from the outset.

Being and Care

Frankly, Heidegger does not have to be the primary philosophical messenger for this position. There are multiple non-Western
sources and philosophies for conceptualizing care.[9] Many thinkers in the existential and continental philosophical traditions have
taken the analysis of intersubjective world-building much further.
And the reasons for critiques of Heidegger's articulation of *Dasein*
are numerous: *Dasein* has no race, no sex, no gender, no parents,
doesn't eat, has no significant insight into the details of biological
conditions, and no ecological commitments beyond Heidegger's
own nostalgia for pastoral lifeways.

The existential examination of life and death also needs to be
supplemented by a biopolitical analysis, theorized by Foucault as the
"matter of taking control of life and the biological processes of man-
as-species."[10] Biopolitical theory stands in critique of traditional
existentialism that validated an ahistorical and universally appli-
cable "lived experience" and "being-toward-death" that presumed
an unquestioned and unmarked notion of life and death guides all
existential understanding. Foucault's historicizing critique of life
and species concepts disrupted the existential faith that "life will
find a way." In biopolitics, the enmeshing of biological and political
practices aimed at managing "life itself," some bodies are prioritized
and optimized, while others are marginalized, subjugated, left to

9. For an Indigenous environmental philosophy of care see
Brian Burkhart, *Indigenizing Philosophy through the Land: A Trickster
Methodology for Decolonizing Environmental Ethics and Indigenous Futures*
(East Lansing: University of Michigan, 2019).

10. Michel Foucault, *"Society Must Be Defended": Lectures at the
Collège de France, 1975–76,* trans. David Macey (New York: Picador, 2003),
246–47.

die, or intentionally killed. The intensification of biopolitical interests in maximizing "ideal" bodies and life principles since the nineteenth century have contributed to the establishment of norms of the human as white, able-bodied, economically productive, and sexually reproductive under patriarchal privilege. The way people and animals are biopolitically positioned in the world continues to be based on doctrines of social health, economic usefulness, racialization, and norms of reproductive sex that will produce bodies beneficial to the trajectories of neoliberalism.

There is a better way to begin: with the embodied, entangled, engaged, ecologically interrelated conditions of existence. This is the existential subject of thinkers like Simone de Beauvoir and Frantz Fanon who analyze the specific "lived experiences" of gender and race. This intersubjective and interobjective condition forms the basis for world-building and world-sharing constructed on ontologically embedded capacities for collective care, consent, and reciprocal duties. The built-in conditions of care and precariousness in existential bodies does not mean that all forms of care are good or that defending care as such will provide a fail-safe route to mitigate existential risks and guide us toward favorable existential possibilities. María Puig de la Bellacasa discusses how the reality of care can be both enabling and exhausting, connective and controlling: "To care can feel good; it can also feel awful. It can do good; it can oppress. It's essential character to humans and countless living beings makes it all the more susceptible to convey control. . . . Care means all these things and different things to different people, in different situations. So while ways of caring can be identified, researched, and understood concretely and empirically, care remains ambivalent in significance and ontology."[11] Bellacasa insists care is about finding the "right distance" (5), not a blanket statement that all interdependency is good.

---

11. María Puig de la Bellacasa, *Matters of Care: Speculative Ethics in More Than Human Worlds* (Minneapolis: University of Minnesota Press, 2017), 1.

If care is both built-in yet also ambivalent and caught up in risks of its own, what does the existential view of care add to the need to cognize and confront existential risks? We readily admit that a commitment to existential thought alone will not stop natural or human-made catastrophes, but practices of care and interrelationality, along with technological developments that allow us to live better together, have helped immensely in making catastrophes survivable in the recent past. Thinking care and catastrophe goes together. Since these practices and technologies continue to help in instances like the COVID-19 pandemic, we see no obligation to prioritize transhuman transcendence over our current relational ontologies. We can focus on improving practices of living and dying together without veering toward either salvation or apocalypse. Certainly the risks of state-based or nonstate agents pursuing mass violence, especially with the use of high-tech weaponry and warmaking, remains paramount, but this does not commit us to an endless technological race for defensive weaponry either. We already know that the only way to win an all-out war is not to play. We can incentivize not playing by reinforcing collective decision-making, consent, and hospitality at a planetary level. We still have to invent what this will look like—it may be a kind of planetary government—but such political forms will require new imaginations and new forms of collective address and response.

For this, we need new forms of democracy and participation, rather than changing the demos at an ontological-technological level so "enhanced" people behave better. Ingmar Persson and Julian Savulescu have suggested that men in particular should be subject to "moral bioenhancement" to quell toxic masculinity—though Persson and Savulescu seem to miss the irony in this masculinist attitude toward championing quick top-down techno-fixes rather than really engaging in the feminist work of achieving gender justice in collective, informed, and noncoercive relations.[12]

12. Ingmar Persson and Julian Savulescu, "Getting Moral Enhancement Right: The Desirability of Moral Bioenhancement," *Bioethics* 27, no. 3 (2011): 124–31.

Several thinkers in the existential risk field including Bostrom have suggested that some form of pervasive surveillance, perhaps coupled with a benevolent authoritarianism (human or techno-logical), would be the solution with the most likely chance of success in thwarting dangerous individuals or groups bent on world destruction.[13] Yet it is truly hard to envision any version of these two in combination that would stop all malicious agents, and these technologies and political forms would generate other existential risks. It may be that the only way to truly mitigate, if not perma-nently prevent, existential risks would be to convince everyone to share the planet—sharing both benefits and burdens, risks and rewards, joys and compromises. This kind of radical sharing can take many forms—and the point is that no one single idea or way of existence can "own" the earth. The existential project is the very commitment to expand the space of existence by existential means of care, being in the world, and coflourishing. This project follows the evergreen slogan: respect existence or expect resistance.

Undoing the Existential

To return to Bostrom's work now, this broad definition of "exis-tential" indicates why neither he nor anyone else in the field has bothered to further analyze the term. Existential risk philosophy provides no special insight into the philosophy of subjectivity and is not interested in the existentiality of philosophy itself as meth-odologically tied to precarity and finitude. Quite the opposite is the case, since Bostrom is more interested in smoothing the way toward a future superintelligent existence at cosmological scales than ex-amining current risks and rewards of being a precarious human subject. Bostrom would readily admit his version of existential risk is anthropocentric—for now. Yet his philosophy is predicated on a

13. See Nick Bostrom, "The Vulnerable World Hypothesis," *Global Policy* 10, no. 4 (November 2009): 455–76.

judgmental assessment of humanity's limitations. Bostrom claims it will be rational for humans to accede eventually to posthumanity,[14] abandoning anthropocentrism for a new post-anthropic condition. Bostrom offers no reflection on the existential risks of ecosystems, animals, and the planet unless they are relevant to human and later posthuman flourishing. Thus he adds technocentrism to a triumphant but temporary anthropocentrism, stating that "our focus should be on maximizing the chances that we will someday attain technological maturity in a way that is not dismally and irremediably flawed."[15] Technological "maturity," in part, entails overcoming human existential perishability and fallibility in favor of transhuman superintelligence. It would be an existential failure to remain merely human. In a kind of anachronistic apocalypse, a future superior is invoked based on the abandonment of existentiality deemed to be already futile in advance.

Bostrom's long-term proposals to avoid existential catastrophe amount to making the case to de-existentialize the human. The larger aim is to replace the precarious and disaster-prone condition of existence with a more deterministic, reliably calculated, and intelligent existence. Yet de-existentializing people is deeply suspicious because doing so would echo the kinds of oppression that have marked so many disastrous political projects to destroy peoples. Transhumanism would discard the existential structure, which is premised on the paradox of the necessity of freedom as Sartre continually stressed, in the name of an algorithm that would be even more free: "Transhumanists promote the view that human enhancement technologies should be made widely available, and that individuals should have broad discretion over which of these technologies to apply to themselves (morphological freedom), and that parents should normally get to decide which reproductive tech-

14. Nick Bostrom, "Why I Want to Be a Posthuman When I Grow Up," in *Medical Enhancement and Posthumanity,* ed. Bert Gordijn and Ruth Chadwick (N.p.: Springer, 2008): 107–37.

15. Bostrom, "Existential Risk Prevention," 25.

nologies to use when having children (reproductive freedom)."[16] Yet actually existing "reproductive freedom" (which includes "cyberfeminist" and trans freedoms to "hack" the body and redesign it toward gender abolitionism) does not necessitate transhuman technology. And "widely available" is not the same as consensual, free, and universal. Transhumanism draws from the historical language of freedom when convenient, but this movement makes no attempt to ally itself with the longer history of social justice movements.

Curiously, a kind of rudimentary existentialism blossoms among the motives of many transhumanists. In an informal poll of adherents taken by Anders Sandberg, he finds many have turned to transhumanism as a pathway to finding meaning in life that they identify as the core existential project. Sandberg notes that "many of the respondents were clearly existentialist in outlook."[17] Hence Sandberg warms to the proposition that, "Transhumanism might be what enables us to lead truly meaningful lives in a physical universe" (8). Here transhumanism is both a rejection and a fulfillment of existential thought. Transhumanism dreams of total self-determination and self-rationalization and confuses this with an un-conflicted existentialism reduced to egocentric self-mythologizing, what Peter Gordon calls an "ontology of wish fulfillment."[18] Transhumanism declares itself more existential than existentialism. It privileges the longevity of egoic selves enhanced by technology and looks condescendingly at the core existential conditions of anxiety, desire, and care that are wrapped up in finitude. Lévinas's insight that "a being without anxiety would be an infinite being—but that concept is self-contradictory,"[19] applies to all aspects of the existential condition.

16. Bostrom, "In Defense of Posthuman Dignity," 203.

17. Anders Sandberg, "Transhumanism and the Meaning of Life," in Mercer and Trothen, *Religion and Transhumanism*, 4.

18. Peter Gordon, *Adorno and Existence* (Cambridge, Mass.: Harvard University Press, 2016), 149.

19. Emmanuel Lévinas, *From Existence to Existents*, trans. Alphonso Lingis (The Hague: Martinus Nijhoff, 1978), 20.

The attempt to use the existential condition to escape the existential condition appears contradictory and self-defeating.

Transhumanism would exchange existential risk for indefinite existential security, foregoing our currently limited capacity for self-determinacy and relationality for the sake of a superior determined artificial superintelligence. As this superintelligence would show us the way to ever more and more intelligent solutions to any problem of immediate risk, nothing we would define as existential would be relevant anymore. No existential precarity would also mean no more ecology—the superintelligence could make or unmake species and ecosystems as needed. Problems of existence and ecology would be replaced by programming problems. However, scientists have pointed to limits of complexity that computation and the "entailing laws" of mechanical determinism can never cross.[20] The question remains why transhumanism should be desirable and what kind of violence, to human and nonhuman life, will result from the side effects of failed efforts to achieve it. Here we can point to the now evident irony in the title of Bostrom's "Future of Humanity Institute"—the future of humanity is not to be human anymore.

Superintelligent Values

Now, quite evidently, Bostrom is reasonably wary of the transformative possibilities of all current artificial intelligence projects. In *Superintelligence,* he has written lengthy and astute analyses questioning how one might encode human ethics into an intelligent machine, how we might ensure that the superintelligence follows these ethical rules, and how problematic or unexpected issues might arise even if the first two matters were achieved. Even if it is possible to encode human ethics into a machine (the "value alignment problem"), there is not a perfect consensus on the universal standards,

20. See John Horgan, *The End of Science* (New York: Broadway Books, 1996), and Stuart Kauffman, *Humanity in a Creative Universe* (New York: Oxford University Press, 2016).

objectives, and methods of morality among humans. Humans often lack consistent value alignment with each other—a problem consistent with the existential condition that lacks metaphysical guarantees. That values are variable and contradictory across many axes of identity and experience is fundamental to the very distinction between fact and value. Furthermore, even if ethics are encodable, a superintelligent machine might choose to rewrite its own code to favor itself. Such a machine may no longer be fixable or controllable by its makers, which is the very principle of self-programming that would mark the ability for AI to truly learn and evolve.

Even good people with good intentions may end up effectuating immensely bad outcomes as technological superpower exceeds traditional forms of control. In yet another scenario, a superintelligence might intentionally or unintentionally cause human ruin by focusing narrowly on one objective and destroying the world to achieve such ends. This problem is emblematized in Bostrom's example of the paper clip scenario: an AI machine is tasked with maximizing paper clip manufacturing and decides to turn all materials on earth, including humans, into resources for paper clips. Such scenarios may of course be unlikely or impossible (looking back to chapter 2, we mark the undertheorized difference between unlikely and impossible as a fundamental feature of existential risk). Yet the idea of a factory dead set on cornering the cosmic market for paper clips is a capitalist fever dream—who would code a machine to make endless paper clips in the first place?

A universe of paper clips aside, many in the AI-risk community emphasize that the most realistic concern is that an extremely competent superintelligence may not end up being malevolent, but just doesn't care about humans. The superintelligence will treat humans as inconsequential or just more data to be managed or ignored. It also bears mentioning that Bostrom and others offer no reflection on the lack of value alignment humans have with other life on the planet—evidently value alignment with machines has a higher priority than value alignment with animals. Another problem occurs in the situation that there may be no way to align existential human values with a nonexistential or postexistential entity. Hubert Dreyfus argued some time ago that computers will never achieve artificial general intelligence without an existential sense of being-in-the-world and with an intuition of horizons of possibility relevant to that situation.[21] Finally, an AI may outrun the existential condition entirely: consider the situation of chess programs unbeatable by humans today. If the machine algorithm always wins, is there really a game anymore? The existential qualitative activity of play, in which the players cannot anticipate all the possible moves even of a "solved" game, can be de-existentialized in a program that always wins, thus in effect annulling the game as a game.

There never has been a situation in history in which everyone has had equal access to a new technology. The makers of superintelligence may want it to be available to all—note that this is not the same as democratically made—however, the technology itself may stand in the way of equality of access. In the case of superintelligence, Bostrom points out that the first person or group who achieves such technology may be able quickly to dominate the entire world. One might have said the same thing about nuclear weapons. This kind of radical, nondemocratic power may also be achieved in more limited yet still transformative technologies including those focused on longevity, de-extinction, consciousness simulation, or biotechnological enhancement. Advances in these fields could involve a situation in which some humans have a kind of technological alignment that is not just temporarily superior to others but can lead to permanently overwhelming and dominating the entire field. The hopeful fantasy of "widely available" transhuman technology may not come to pass, especially given that near-term AI work is driven almost entirely by capitalist corporate entities that are not beholden to public egalitarian goods.[22] The fantasy of endless expansion of

21. Hubert Dreyfus, *What Computers Can't Do: A Critique of Artificial Reason* (Cambridge, Mass.: MIT Press, 1972).

22. See Nick Dyer-Witheford, Atle Mikkola Kjøsen, and James Steinhof, *Inhuman Power: Artificial Intelligence and the Future of Capitalism* (London: Pluto Press, 2019).

superintelligence is modeled on the endless expansion of capital and self-interested economic growth, not sustainable social communion.

Despite all the additional existential risks posed by this kind of technology, Bostrom readily asserts that superintelligence is too promising not to develop. In the words of an early computer scientist, I. J. Good, "The first ultraintelligent machine is the last invention that man need ever make, provided that the machine is docile enough to tell us how to keep it under control."[23] Superintelligence could possibly solve our planetary ecological problems, maximize the capture of solar energy, manage the climate, produce ethical food, develop de-extinction sciences, cure our diseases, and solve our philosophical problems to boot. This of course would just be the beginning. For Bostrom, the greatest existential risk is to remain existential, which is by definition risky.

Left-Wing Prometheanism

Even if some of these technological and existential turning points are decades or centuries away, decisions already are being made today in existential risk policy, AI development, and sciences of longevity that follow along the lines that Bostrom and others in the field have set forth. A robust critique of existential risk is needed now. A few important analyses of existential risk from the perspective of critical theory have taken on some of Bostrom's claims. Claire Colebrook, in her essay "Lives Worth Living: Extinction, Persons, Disability," slams Bostrom's theory as perpetuating "a logic of appropriation, extension, survival, and calculus" that perpetuates widespread dominations.[24] Colebrook especially is cogent in discerning how Bostrom's philosophy elevates intelligence as the highest value by which all life shall be compared and judged. Bostrom never really

23. Bostrom, *Superintelligence*, 4.

24. Claire Colebrook, "Lives Worth Living: Extinction, Persons, Disability," in *After Extinction*, ed. Richard Grusin (Minneapolis: University of Minnesota Press, 2018), 168.

defines "intelligence" (though he does put some stock in IQ tests and other questionable metrics)—what is more important is that measurable quantifications of intelligence, be they human or artificial, continue to show a rise that will lead toward technological maturity and enhanced forms of living and away from existential dangers. As Colebrook discusses, the privileging of the abstract category of "intelligence" as the barometer of the "future of humanity" does not offer any reflection on the long history of violence and expropriation by those deemed possessed of superior intelligence and capabilities. Colebrook argues that "scenarios of catastrophic risk—such as those of Nick Bostrom— . . . assume that humanity is necessarily defined by a certain concept of personhood that is irreducible to the human species. Indeed, it is ability—in Bostrom's case, intelligence—that needs to be preserved; it is *this* life that would count as extinction and not 'merely' as genocide" (170). Colebrook discerns how Bostrom sets up intelligence as what Sylvia Wynter calls a "new master code"[25] that elevates some humans over others and eventually elevates transhumans over humans. Colebrook sees in Bostrom's utilitarian framework a permission to calculate human lives according to hierarchies of intelligence that have "the logic of extinction" (167) at their core because they inevitably posit that some lives are to be distinguished as good lives (more intelligent) while others are deemed less worthy.

Colebrook, in her recent writings, includes in her summary critical judgment of humanisms, posthumanisms, and transhumanisms any thought that campaigns for indefinite survival and disregards the non-humanist forces of matter that currently make up ourselves and our world and that continue to effectuate geological transformations that do not prioritize human standing or self-interests.[26] Even the existential condition and its attachment

25. Sylvia Wynter, "Unsettling the Coloniality of Being/Power/Truth/Freedom: Towards the Human, After Man, Its Overrepresentation—An Argument," *CR: The New Centennial Review* 3, no. 3 (Fall 2003): 323.

26. See Claire Colebrook and Jami Weinstein, "Preface: Postscript on

to worlding and care are found to be suspect in Colebrook's vision because they harbor a built-in exclusionary ethos such that, to use Heidegger's language, authentic humans have worlds to care about, animals (and perhaps inauthentic humans) are poor in world, and stones are worldless. Accordingly, only a truly inhuman or outside-of-humanist perspective, unsentimental about any yearning for salvation or possessive worldliness, can circumvent the apocalyptic drive of the human or posthuman that aims to persist at any cost.

At the same time, and perhaps not fully avowed, Colebrook's thinking and writing does not wholly avoid the effects of calculative valuation and persistence. No form of reasoning or being in the world is fully distanced from some use of calculative and persistive rationality. It is possible to reject the sovereignty of technoscientific reason and calculative thought while still employing these tools in the very disassembling of exclusionary hierarchies of intelligence. We should be able to use calculative reason in assessing existential risks, and still criticize the reliance on statistical modeling, so that we are not oppressed by the sheer pursuit of more and more technoscientific success. Moreover, we need not concede that intelligence is what Bostrom says it is; multiple forms of intelligence (ecological, emotional, ethical intelligences) refuse the triumph of superintelligence. It is possible to critique concepts of existence, intelligence, and the good life while still employing them, striving to improve upon the implications of the human/nonhuman entanglements of a shared world, while not denying the fragility and entropic tendencies of matter and meaning. Furthermore, if one does want to salvage durable institutions predicated on collective justice, these must rely on calculating situations of rights and reciprocities while continually raising critiques about the limits and problems of calculative reason.

the Posthuman," in *Posthumous Life: Theorizing beyond the Posthuman,* ed. Colebrook and Weinstein (New York: Columbia University Press, 2017), 1–16.

Is it just nostalgic to want to keep our current existential structure? Are we clinging to what Ray Brassier calls "unobjectifiable transcendence,"[27] whereby we deem the current existential condition untouchable and transcendent by the sheer fact that this condition is what we have always had? Is the existential–ecological condition founded and legitimized by a presumed traditionalism or naturalism or protected by some other "myth of the given"? Brassier, from a very different political and philosophical approach, ends up largely agreeing with Bostrom as to the principled ends of using technology to de-existentialize humans. Brassier, whose philosophical commitments include analytic traditions of scientific realism, continental theory, and Marxism, advocates a "Left Prometheanism" (against what he sees as Bostrom's Prometheanism of the Right) that would allow us to remake our own current given existential conditions in order to achieve a more just world. Brassier disputes that the existential structure characterized by physical and cognitive finitudes is to be treated as a transcendental condition that we are passively "thrown" into. He questions the existential argument that humans would be unwise to disturb a supposed balance between the given (our initial human condition) and the made (what we do with the human condition to make it meaningful). "I take this claim that we ought to respect the 'fragile equilibrium' between what is made and what is given to be fundamental for the philosophical critique of Prometheanism" (474).

In *Nihil Unbound,* Brassier picks apart arguments in the existential tradition that the human's orientation toward their own death provides the privileged horizon generative of care and interpretive

27. Ray Brassier, "Prometheanism and Its Critics," in *#Accelerate: The Accelerationist Reader,* ed. Robin Mackay and Armen Avanessian (Falmouth, UK: Urbanomic, 2014), 477. For a critique of the radical Left techno-politics of accelerationism, see Benjamin Noys, *Malign Velocity: Accelerationism and Capitalism* (Alresford, UK: Zero Books, 2014). For "middle-ground" between the two positions, see Steven Shaviro, *No Speed Limit: Three Essays on Accelerationism* (Minneapolis: University of Minnesota Press, 2015).

meaning.[28] The fact of humanity's inevitable extinction is crushing evidence that death does not provide a horizon of meaning and purpose but precisely indicates the opposite, the "realist thesis" (235) that the metaphysics of purpose does not escape physical entropy and that an "originary purposelessness . . . compels all purposefulness" (236). Extinction is a kind of objective, physical realism that annuls human investment in subjective transcendentals as anything more than temporary benefits. "Extinction is not to be understood here as the termination of a biological species, but rather as that which levels the transcendence ascribed to the human, whether it be that of consciousness or *Dasein*" (224).

Heidegger and others in the existential tradition construe the special relation of humans toward their own horizon of meaningfulness as sufficient to claim that it is finitude and precarious life that structure humanity's essential condition. Yet Heidegger overstates the specialness of human autonomy, which is composed from the outset by heteronomous relations with symbiotic animals (including the microbiome living inside each person), tools and machines (including language), and inanimate geological materials and forces. For Brassier, there is no real "prohibition on self-objectivation"[29] that forbids humans to reengineer themselves, since there is no transcendental status or "fragile equilibrium" between givens and mades. Our lives are already inextricably tied to instrumental, engineered, and cybernetic conditions, so the point is now to conduct these more rationally and equitably. If there is nothing special that upholds this existential condition predicated on claiming its deepest insights from its own finitude, why not change it? Brassier asks pointedly, "What exactly is reasonable about accepting birth, suffering, and death as ineluctable facts, which is to say, givens?" (479).

From a very different perspective than those of the existential risk community, but consistent with their posthuman logic, Brassier

28. Brassier, *Nihil Unbound*, ch. 6.
29. Brassier, "Prometheanism," 476.

urges the reconfiguration of the human according to the rational pursuit of collective goods such as social justice, the overcoming of suffering, freedom from economic burdens and scarcities, and shared technological progress. Brassier also adds that the pursuit of reason—which for him entails the making and following of rational, equitable rules as well as the development of scientific reason and logic—warrants technological as well as philosophical means to accomplish a more rational world. Reason formally compels its own pursuit such that humans are able to revise their concepts and ways of knowing in order to pursue a more correct and just picture of reality. Yet this apotheosis of reason is somewhat in contradiction to the human-centric Left Prometheanism Brassier espouses, since in his book on extinction he adamantly claims that reason cannot serve as consolation for the finitude of life. Humans should pursue the progress of reason, but reason itself is not human and is not bound to the cares and exigencies of the lifeworld.

Yet, to apply Brassier's own commitment to nihilism to his project, a nihilism that for other thinkers provides an important gateway to existential thought, this pursuit of reason cannot serve as a new necessary metaphysical or transcendental project. Humans are not obliged to technologically change themselves to meet the standards of a relentless and disconsolate rationality. The Left Promethean project remains anthropocentric and relatively unconcerned about the continued domination and destruction of terrestrial nature. There is no overarching "reason" that strictly privileges human interests over all sentient beings, nor brings life into conformity with logic. Just because the givens and the mades of life are not in any transcendental equilibrium does not mean that they should all be revisable and perfectible, and especially revisable according to human self-interests. It also remains questionable how Brassier's commitment to a realism of purposeless materiality should apply to the project of planetary rationalization, which depends on the assumption that ecosocial systems answer to logical ordering.

Indeed, it may be rational for humans to forgo such a claim for control over everything on the planet, given that the unleashing of

Promethean powers to change the human condition would certainly also unleash powers to destroy everything on the planet as well. We need not genuflect to the belief that natural existence should be pristine and never altered, but this does not mean nature should just do our bidding, or even that we should save nature from the "irrational" woes of mortality. Brassier's genuine exhortation to pursue a communist Prometheanism neglects that these very same technologies to remake the existential condition may end up exceeding any form of reason or pursuit of collective good, and may no longer be revisable due to irreversible technological circumstances. Such powers over life may not be handled reasonably by the collective (as in, for example, the claims for species privilege most humans currently assert over nonhuman animals) and may preclude collective revisability because such technologies may evolve beyond human control or rational cognition. In pursuit of the exigencies of reason, we may make ourselves unfixable and uncontrollable and unreasonable. The existential capacity to revise the given and the made itself must not be made into something unrevisable. The Marxist project has elements of Prometheanism and elements of ecological care that combine in the work of environmental and social justice—there is no need (and perhaps no way) to reduce one to the other, and while there may be no "equilibrium" between the two, there can be a dialectic, a reciprocal and dynamic interchange.

Cosmic Calamity Theory

To return to Brassier's commentary on "humanity's inevitable extinction," we would be remiss not to look at his argument about the death of the sun and compare it to the buffet of extinctions offered up in the field of existential risk. The eventual transformation of the sun into a "red giant" and then diminishment into a "white dwarf" as it completes its cycle of thermonuclear reactions presents a contrast to "risks" such as AI, climate change, and asteroids. Unlike them, the sun's demise is inevitable. In the final chapter of *Nihil Unbound,* Brassier draws on nonconsoling thinkers such as

Nietzsche and especially Jean-François Lyotard's argument about "solar catastrophe" in his book *The Inhuman*. Lyotard divulges a hyperbolic attitude about the retroactive negation of meaning, because this point comes in a dialogue between the characters "He" and "She," where "He" exaggerates his point with a staccato of sentence fragments:

> Human death is included in the life of the human mind. Solar death implies an irreparably exclusive disjunction between death and thought. If there's death, then there's no thought. No self to make sense of it. Pure event. Disaster. All the events and disasters we're familiar with and try to think of will end up as no more than pale simulacra.[30]

This end of thought that thought cannot possibly think erases the horizon. The male voice in the dialogue argues that this *ultimate* question is "the sole serious question facing humanity today" (9). In valuations that recall Bostrom's typology of risks and privileging of the kind that would lead to the extinction of "Earth-originating intelligent life," Lyotard (or the character "He") goes on to say that "wars, conflicts, political tensions, shifts in opinion, philosophical debates, even passions"—all are "dead already" because "the explosion to come . . . can be seen in a certain ways as coming before the fact" to render the list of attachments "posthumous" and "futile" (9). Brassier's point about the death of the sun is similar, though his Prometheanism is very different from Lyotard's suspicion of technoscience.

What these thinkers share most in common is the logic of retroactive negation grounded in the thought of deep-future extinction. For Brassier, this means that *posteriority* disqualifies the importance of subject–object correlations crucial to existentialism much like Quentin Meillasoux's concept of ancestrality does from the vantage of the past. Deep-future extinction is thermodynamic, made inev-

30. Jean-François Lyotard, *The Inhuman: Reflections on Time*, trans. Geoffrey Bennington and Rachel Bowlby (Stanford: Stanford University Press, 1991), 11.

itable by the law of entropy: potential energy exhausts itself. Even if humans somehow escape the solar system, Brassier reminds us, this would only defer the inevitable:

> sooner or later both life and mind will have to reckon with the disintegration of the ultimate horizon, when, roughly one trillion, trillion, trillion (10^{1728}) years from now, the accelerating expansion of the universe will have disintegrated the fabric of matter itself, terminating the possibility of embodiment. Every star in the universe will have burnt out, plunging the cosmos into a state of absolute darkness and leaving behind nothing but spent husks of collapsed matter.[31]

With this point about cosmological exhaustion and darkness, Brassier mocks that no deferral through transhumanism or becoming-interstellar will save humans from extinction. The concept of posteriority is less a posteriori or inductive logic, more retroactivity before the fact. Because "we" humans and all of life will inevitably be extinct, there is *nothing now*. As Brassier puts it, "the *extinction of spacetime*" "already cancels" the correlation of being and meaning, with existentiality "already retroactively terminated" (230).

In chapter 2, we showed how this future anterior logic works for existential risk theorists such as Toby Ord. But there is also a clear contrast between their conception of unprecedented events and Brassier's ideas about extinction. In fact, the two ways of thinking are almost opposing limit cases: extinction is either inevitable or radically contingent, determined in advance or an array of possible ends of human life, unavoidable or something that we should strive to avoid until humanity can achieve the full maturity of superintelligence. Humans (or Earth-originating intelligent life) are either potentially godlike or "spent husks" of nothingness. Through this opposition, Lyotard and Brassier's own work can serve as a critique of the theological metaphysics of existential risk and a return to the reality of finitude. In *The Inhuman,* Lyotard was already working against the idea that thought can go on without a body, and his argument links transhumanism to the inevitability of extinction as

31. Brassier, *Nihil Unbound,* 228.

early as the 1980s. In their effort to mitigate all future extinction scenarios, existential risk theorists seem to ignore the inevitable, or to surreptitiously set the time frame of concern *back* to a time arbitrarily nearer to us in the lifetime of the universe. Even transhuman minds will have to "reckon with the disintegration of the ultimate horizon" (228), which may or may not preclude concerns about less encompassing extinctions that could happen in the trillions of years before the end of all ends. Even if heat death functions as a critique of existential risk, however, the more subtle reason to read Brassier here is to look for the third term (expanding the space of existences across multiple temporal and cosmological conceptions) in this opposition. By way of questioning heat death, we work toward conclusions about how existential ecology should relate to cosmic timescales.

Taking up the theme that has struck fiction writers from Camille Flammarion to Philip K. Dick, Brassier makes entropy the foundation for retroactive negation. How striking that, on the one hand, the death of the sun and the heat death of the universe seem ineluctable and, on the other hand, that theorists could notice this but close their concepts off from key cosmological questions. The combination of entropy and rationalization of finitude into the terms of nothingness, however, raises the question of how empirically sound is the long-standing cosmic inflation hypothesis, with its reliance on the inevitability of heat death.

Within cosmology, the Big Bang and cosmic inflation have faced challenges in recent years. We do not need to detail the conflicts between the architects of the Big Bang theory and cosmologists like Paul Steinhardt and Neil Turok, who argue that observational evidence is more consistent with infinite, cyclic, or bounding models of a universe that is not unified and lacks a beginning or end.[32] Historians of science like Bjørn Ekeberg have suggested that, since the late twentieth century, cosmology and theoretical physics

---

32. See, for example, P. J. Steinhardt and N. Turok, "A Cyclic Model of the Universe," *Science* 296, no. 5572 (2002): 1436–39.

have gone too far in trying to force their observations into rational (mathematically unified) models of the universe, inventing new concepts that *must* be real even when they cannot be observed.[33] So it is easy to cast doubt, from a scientific perspective, on Brassier's claim of ontological disenchantment. What is striking about his repetition of heat death and inflation theory from only two scientific references is that a theory that depends so much on having the right science about the behavior of matter at the largest scales, in an unobservable future about which we can have no empirical evidence, needs its science to be not just factual but *absolute*.[34] As is often the case when thinkers choose a science to make absolute, much of the actually existing debate and complexity around a theory like inflation vanishes.[35] It is dubious to see such remote cosmic events treated as the hard truth of reason and materiality, not only because cosmology is a rather fuzzy and speculative science on which to stake one's bets, but also because the thought of the end of the universe is so wonderful and rich in questions. The single most common ploy of scientism is to present some still-ambiguous topic as unproblematically true.

These scientific questions raise the usual problems of contingent evidence, the absoluteness of laws, ever-changing observations of the universe, and the technologies that enable them. Yet none of these empirical problems quite capture what is most amazing about Brassier's argument for disenchantment, which is that it forgets questions of being at the exact moment where they seem unavoidable: the end (or the beginning, for that matter) of the universe. Such questions are as childlike as they are important: What came

33. Bjørn Ekeberg, *Metaphysical Experiments: Physics and the Invention of the Universe* (Minneapolis: University of Minnesota Press, 2019); Bjørn Ekeberg, "Cosmology Has Some Big Problems," *Scientific American*, April 30, 2019, https://blogs.scientificamerican.com/observations /cosmology-has-some-big-problems/.
34. Brassier, *Nihil Unbound*, 203.
35. To be fair, much of the popularized controversy about cosmic inflation is more recent than Brassier's book, if not the specialist publications.

before the beginning and what will come after the end? What is the being in which the material determinations studied by cosmology and astrophysics play out? What lies beyond the outer frame of our secular scientific worldview? Are there cosmological "ends" at all? If "ends" are an open question, where does this leave the cosmological question of extinction?

In fact, these questions could be asked at any stage of cosmic evolution, but it's the concepts of origins and ends that make them most urgent when referred to extinction horizons in the present, while also raising our deconstructive suspicions. Without the expertise to decide one way or another, but keeping in mind the current debates in cosmology, would it be perverse to be suspicious about the metaphysics of a Big Bang theory that has such a clear narrative structure of beginning, middle, and end? Even if the theory is correct, this might not reduce away the mystery of the broader context in which the universe is evolving from Bang to entropic Whimper. The radical mystery of this context should not be seen as an answer to human finitude or an escape from extinction, which is the predictably transhuman role played by the "fifth dimension" in Christopher Nolan's film *Interstellar*. But this persistent question of being does seem to militate against Brassier's aggressive interpretation of heat death as disenchanting and demystifying, revealing the ultimate nullity of ontology. Questions about what lies before the beginning or after the end are spurs toward the kind of speculative cosmology that might inform alternatives to risk management and transhumanism.

Can Utopia Be Compulsory?

Cosmological frames of analysis like Brassier's and Bostrom's "astronomical" value horizon treat the ultimate ends of evolutionary processes as having conceptual and methodological primacy over any other point in time, including the present. Selecting for scale and a specific "stage" (if it still makes sense to talk of beginnings, middles, and ends) in any evolutionary-historical process always will have outsized effects on the ontological claims of any theory. Our preference for existential ecology has its own framing problems, although the advantage of this scale is that it can process the insights of cosmological entropy as already constitutive in the formations of planetary ecologies and the existences that share these worlds in reckoning with their own finitudes. The cosmological scale of heat death or the astronomical scale of transhuman superintelligence lack a robust theorization of their own implications in ecological and existential precarity in the present. Two more contributions to the tradition of existential thought then can serve as prescient guides for thinking existential ecology in a time of existential risks. The work of Hans Jonas, a Jewish student of Heidegger's at one time, provides a remarkable example of a restatement of existential thought after witnessing the horrors of World War II. Jonas's first major writings on existential philosophy in the 1950s culminated in his volume *The Phenomenon of Life,* which set out to offer "an 'existential' interpretation of biological facts."[36] Jonas claims that "biology turns into ethics" (2) since over the course of biological evolution humans have gained the capacity for different forms of care going beyond physical satisfactions. This is not a strict naturalism but an argument that evolutionary processes, over time, have created the conditions for humans to bear a more substantial capacity of responsibility and carefulness that are the hallmarks of the existential structure. Jonas contends that this natural and existential basis for care has now produced a new *ought*: humans are responsible to maintain the capacity for responsibility itself. Simply put, humans ought to maintain the ability to have oughts. Ethics now includes duties toward the ought itself and to not let perish the capacity for adherence to oughts.

Jonas expands the implications of this argument in *The Imperative of Responsibility,* first published in German in 1979, a lucid book on

36. Hans Jonas, *The Phenomenon of Life: Toward a Philosophical Biology* (Evanston, Ill.: Northwestern University Press, 2001), xxiii.

the state of ethics after technologies of nuclear war, environmental destruction, and genetic engineering. This book now reads as an existentialist critique of the transhuman position on existential risk.[37] Jonas begins by recognizing that the condition of ethics itself has fundamentally changed in the past century due to the apocalyptic power of new technologies. Previous ethical philosophies, from classical Greek philosophy to Kant's categorical imperative, assumed that "the human condition, determined by the nature of man and the nature of things, was given once and for all; that the human good on that basis was readily determinable; and that the range of human action and therefore responsibility was narrowly circumscribed" (1). Now that these premises no longer hold, with human action capable of changing or destroying the human condition as well as the natural condition, morality has changed. Humans are obligated to care for the future effects of their actions and for the future of the plenitude of life on the planet. Humans have an ethical injunction to bequeath the capacity of having such duties to future humans. We do not have an ethical imperative to achieve perfection or dissolve suffering. There is no metaphysical principle that says we must maximize every human potential or wish for ourselves or for future generations. Toward future generations, Jonas argues, we have a responsibility not to grant them "astronomical" benefits but simply to bequeath a future that includes them in it: "In the final analysis we consult not our successors' *wishes* (which can be of our own making) but rather the 'ought' that stands above both of us" (41).

With the expansion of the scope of responsibility, Jonas proposes new categorical imperatives: "That there *ought* to be through all future time such a world fit for human habitation, and that it ought in all future time to be inhabited by a mankind worthy of the human

<hr />

37. Hans Jonas, *The Imperative of Responsibility: In Search of an Ethics for the Technological Age* (Chicago: University of Chicago Press, 1984). Bostrom briefly cites Jonas in the essay "In Defense of Posthuman Dignity" but dismisses Jonas's work as committed to "bioconservatism" (211).

name, will be readily affirmed as a general axiom" (10). There is also the reverse corollary: "Do not compromise the conditions for an indefinite continuation of humanity on earth" (11). These imperatives are meant not to bolster anthropocentric humanism but to restrict humans from putting the whole of humanity and the planet at risk.

Jonas, not a "bioconservative" lamenting technology as such, grants that we live in a technologically mediated world. He stresses caution, care, justice, and reason as practices that can benefit from technology but do not require it, but he adds these commitments are principally and logically against the cult of technology. Jonas especially casts a wary eye to what he calls the "inherently 'utopian' drift" of technology that is channeling humans toward "unwanted, built-in, automatic utopianism" (21). Can utopia ever be compulsory? In Jonas's view, there is no inherent human duty to achieve utopia or immortality or an end to all work and all suffering. Humans are not responsible for perfecting themselves or their world. Ethics is the granting of care to the precarious, not granting perfection to the imperfect. As Jonas states, "Only for the changeable and the perishable can one be responsible" (125). And while there is no inherent obligation to forever keep humans unchanged, Jonas's argument also calls for a ban on making permanent existential decisions for future generations or depriving them of having any access to the existential condition. This duty applies to nonhuman animals as well.

In her book *Wild Dog Dreaming: Love and Extinction* (2011), Deborah Bird Rose defines existential ecology as follows: "Ecological existentialism thus proposes a kinship of becoming: no telos, no *deus ex machina* to rescue us, no clockwork to keep us ticking along; and on the other hand, the rich plenitude, with all its joys and hazards, of our entanglement in the place, time, and multispecies complexities of life on Earth."[38] Rose's "ecological existentialism," articulated in the context of Indigenous-settler reconciliation work in Australia, understands entanglement, a different metaphor than the plateau,

---

38. Rose, *Wild Dog Dreaming*, 44.

as ambivalent as well as fundamentally biodynamic (not "biocon-servative"). This ecological existentialism is immediately accessible as a multispecies commons and not predicated on trillion-dollar technological industries. Such existentialism is not an anthropo-centric humanism but a devotion to explore and experiment within existential relations intertwined with the conditions of biodiversity and earth systems. Rose's existential ecology welcomes not the overcoming of the human but a process committed to constantly redefining the human without the guarantee of rescue or escape from this condition.

In addition to all the ecological existential risks we face in the next fifty years, we face this risk of irrevocably changing all of our planet's existential structures. There is no necessary reason for us to ever affix ourselves and our planet to some superior determinative intelligence, but this does not mean we cannot welcome some ways of existing with some kinds of superintelligence, and perhaps these could even expand, rather than circumvent, the conditions for care in a precarious world. With Frédéric Neyrat, we embrace the idea that Earth is a "traject" rather than an object (a thing viewed from space) or a subject (Gaia, a giant organism or person). The Earth as *traject* is an "interval spanning space time," "a long term event that began 4.54 billion years ago, the historical trajectory of an entity that will disappear in several billion years."[39] The "unconstructable" Earth is an event that we cannot "reproduce in a laboratory"—it can neither be "controlled nor dominated" because it is an unfolding duration rather than a thing that might be broken down into its parts and remade (171). Even if we were to inhabit other planets, we would be extending the traject elsewhere rather than transcending it.

Cosmology and deep time are not distant from today's most ur-gent theory and practice. It seems strange, then, that there is so little attention to cosmological questions in critical theory. In this respect, a critical approach to existential risk must distinguish be-

---

39. Frédéric Neyrat, *The Unconstructable Earth: An Ecology of Separation* (New York: Fordham University Press, 2019), 171.

tween the kind of near-term care that we might hope will help different collectives of humans and nonhumans through events like the 2020 pandemic and the kind of existential thought that is fundamentally speculative, reaching forward to ask whether the final horizon for the existence of life is the death of the sun or the collapse of the Milky Way into the black hole at its center, or in fact something relatively close to us in geological time. We should be curious about the metaphysical and cosmological questions of how humanity, "Earth-originating intelligent life," life, or even being will come to an end. But existential thinking about extinction should keep in mind the difference between shorter timescales at which we can conceivably plan ways of caring for each other and deep time thinking that has a very distinct existential meaning. It might almost be better to choose arbitrary, midlength timescales to frame political theories, like the one hundred thousand years of Madsen's *Into Eternity*, rather than working with ahistorical timescales assumed by goals like sustaining the "future of life" or "safeguarding humanity." This would be a departure from Jonas's norm that there "*ought* to be through all future time such a world fit for human habitation." One must also think the existential meaning of catastrophes such as the current pandemic along with the growing miseries of global heating, which would not count for Bostrom as properly existential risks, and which do not convert into an existential guarantee of a future for humanity. But we are against extending a model of risk management and utilitarianism to truly cosmic scales at which they become absurd ways of denying the inevitability of extinction and thus of blocking those who embrace this model from thoughtfully integrating extinction into their coexistential structures. If the negative critique of existential risk is to roll over into something affirmative, one direction it might take is to understand extinction as the starting point for making new ontologies and mystical practices from the counterintuitive data of physics and cosmology.

Conclusion: Opening the "Letter from Utopia"

AMONG UTOPIAN SCENARIOS realistic or imaginary, how does one rank Bostrom's call for transhumanism as the resolution of existential risk? Is transhumanism's possible "astronomical" value the most promising vision of utopia conceivable? What alternatives arise in response to this case for an astronomical-scale utopia?

We conclude with a closer examination of Bostrom's utopian imagination, in particular his curious "Letter from Utopia" (2008). As we have seen, Bostrom's definitions and methodology of grappling with dystopian existential threats is inextricable from his utopian transhumanism. Every claim, indeed every sentence, in Bostrom's oeuvre is involved in this utopian outlook. Bostrom presents the true solution to existential risks in transhumanism, understood as the culmination of the Enlightenment drive for emancipation and knowledge. This vision also borrows selectively from the fantasy space of Marx and Freud, positing transhumanism as wish-fulfillment, the ego sublime relishing endless surplus leisure.

To develop foresight and anticipate existential risks as well as to build a better world, we certainly need to be mobilizing and expanding our imaginative capabilities. Bostrom has been quite detailed and specific at certain moments about imagining what transhumanist life would be like. Yet these details also betray some deeply problematic and hostile attitudes toward present life and the role of speculative futures. Bostrom habitually relies on different kinds of rhetorical lures, some cast as positive and some negative, to entice readers to not just analyze existential risks, but to want to realize posthuman glories. The title of his essay "Why I Want to Be Posthuman When I Grow Up" divulges how he thinks being a human is immature and childish compared to posthumans.

Bostrom is quick to criticize those who are wary of technological enhancements on the ground of wanting to prioritize human relations and ecological care as "bioconservative," but he often also resorts to naturalist metaphors and analogies to describe posthuman progress. The passage from child to adult is a natural progression (and yet cultural in every way too): "Yet we do not think it is bad for a child to grow up. . . . Why then should it be bad for a person to continue to develop so that one day she matures into a being with posthuman capacities?"[1] Human maturation is biologically necessary and logically undeniable—though definitions of child and grown up vary widely across cultures—while posthumanism (if truly possible) is an option. Yet Bostrom conflates nonoptional maturation with the transition to posthumanism that he keeps maintaining would be optional (the essay title suggests posthumanity is a "want," not compulsory). Here the rhetorical lure of the analogy child-adult/human-posthuman is both positive and negative: posthuman maturity is good, so don't be such a (human) child.

What, then, will these posthuman adults be doing? As the argument shifts in this essay from the first-person "I" to using the direct address of the second-person "you," Bostrom is quite specific in who he thinks "you" will become:

> As you yourself are changing [in the early steps of the posthuman process] you may also begin to change the way you spend your time. Instead of spending four hours each day watching television, you may now prefer to play the saxophone in a jazz band and to have fun working on your first novel. Instead of spending weekends hanging out in the pub with your old buddies talking about football,

1. Bostrom, "Why I Want to Be a Posthuman," 16.

you acquire new friends with whom you can discuss things that now seem to you to be of greater significance than sport. Together with some of these new friends, you set up a local chapter of an international nonprofit to help draw attention to the plight of political prisoners (5).

Beer vs. political rallies, screen time vs. jazz. The class snobbery here is tremendous. People who watch tv or like sports are viewed as childish, lacking upwardly mobile intelligence and upwardly mobile existence. Those who are too physically and mentally exhausted after work to do little more than consume media are to be shunned. So much for loyalty to friends—posthumans don't hang out with "old buddies" in the pub, as socializing for the fun of it seems to be beneath the posthuman condition.

Instead, the posthuman will be pursuing experiences that are seen as directly useful for self-improvement and morally congratulatory. For Bostrom, these apparent upgrades in lifestyle seem to require elaborate enhancement technology, when all they require right now is some creative effort and dedication. Maybe a major in the humanities would help. But Bostrom treats the humanities in general as a lure for posthumanity, not as something enjoyable or enhancing to do right now for its own sake. We will supposedly have more time in posthuman life tomorrow to do the humanistic things we all wanted to do but do not have time for presently. But one can learn saxophone or support Amnesty International right now, no supertechnology needed. Go ahead and write that novel today. Yet there is a heavy whiff of dystopianism in the cold shoulder given to those "old buddies" while "new friends" beckon.

Bostrom's utopian platform requires a pathway to reconcile the real and the ideal; an irresolvable gap between the two would constitute yet another existential threat to realizing utopia's "astronomical" value. For him, utopia cannot remain a thought experiment or mere spur for the imagination but must be technologically attainable as long as there is nothing inherent in technology itself that would limit its realization or precipitate more existential risks. We can use the semiotic square to chart out the variations in the

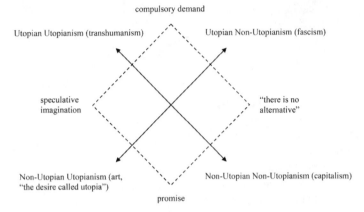

Figure 1. Diagramming Utopianism in the Semiotic Square.

mapping of utopia that can be discerned in the difference between a utopian demand that forces its realization (as in Jonas's compulsory "automatic utopia") and a utopian speculative imagination that is a perpetual promise but can never fully be realized.

Bostrom's transhumanist position leaves no gap between utopia and utopianism, requiring an actualized utopia to satisfy the utopianist demand. This position continually subsumes the speculative imagination into the technological process of actualization, as we will see with Bostrom's "Letter from Utopia," where supposedly all schemas of the imagination can become "lived experience." Fascism, by contrast, posits a fixed imagination (there is one ideal race and one way to achieve its political power) that demands to be realized in an actual utopian state that is premised on nonutopian values. Capitalist realism, the present condition of neoliberalism, maintains the premise that all actually existing utopias as well as all utopian imaginations always collapse into dystopian regimes.[2]

2. Mark Fisher, *Capitalist Realism: Is There No Alternative?* (London: Zero Books, 2009). The question remains where to place communism in this semiotic square: if communism is understood as synthesis, it could be

Since there is no alternative to capitalism in this scheme, the utopian imagination is better commodified or ridiculed.

Non-utopian utopianism (pace Fredric Jameson) is the space of art, the idealizing imagination, and the shaping of political and psychological desires in fantasy and wish-fulfillment. Art posits utopian values that need not be realized or realizable for these to remain essential guides for living in common. While transhuman utopianism insists that redeemed life be directly representable, graspable, and masterable, art flourishes in the gap or delay between representation and reality, a gap that compels both desire and disinterestedness. The utopianism of art's form lies in its very resistance to mastery and ownership, and in the demand for the artwork to be accessed and reinterpreted indefinitely but without any guarantees of meaningfulness. Hence the utopia of art's form is found in the gap itself, the difference between ideality and reality that perpetuates non-utopian utopianism. The existential structure of the artwork refuses fixed totalizations while positioning experience in the realm of "promise" or fantasy, which is not sequestered from sociality but seeps into reality and effectuates everyday changes, as the work's form carries within an incommensurable desire for singular enjoyment and collective reconciliation.

Utopia Writes Back

Bostrom's "Letter from Utopia," perhaps his most fully realized transhumanist scheme, is a fictional letter written from the first-person perspective of a highly enhanced posthuman from the future who is trying to convince humans of today to pursue biological and cognitive augmentations. The letter is written as if postmarked from Utopia itself. It is addressed to "you," the reader. It is signed,

---

placed right in the middle of the X. If communism is considered to maintain an unreconciled and ever-unfulfilled dialectic, it would be between Utopian Utopianism and Non-Utopian Utopianism.

"Yours sincerely, Your Possible Future Self,"[3] suggesting that "you,"
in fact, are the writer of the letter. The letter's explicit purpose is
"to tell you about my life—how good it is—that you may choose it
for yourself" (LU, 1).

There are many paradoxes at work in this genre of letter. The
writing of this letter in the present is what will create this future
writer—the future writer does not exist unless the letter is written
and finds its proper reader in the present. The "Possible Future
Self" is a ghost from the future who, if effective, would haunt himself into becoming an actual future self. The present address thus
proleptically causes the future writer to exist, but the future is also
retro-causally acting in the present. Bostrom, who stands in for the
"we" of present humanity, is both author and reader, "I" and "you."
Bostrom's letter writer is both the addresser and addressee, sender
and receiver, present and future, haunter and haunted by the transhumanist call, writing toward the future and imagining the future
writing back. By collapsing these distinctions, he intends his letter
to close the gap between fantasy and reality.

The letter is an "invitation" (LU, 1) to "your actual future" (LU,
1). What will this future entail? Everything is bursting with "bliss"
(LU, 2). We are asked to remember our best moments and feelings,
and then told "what you had in your best moment is not close to
what I have now—a beckoning scintilla at most" (LU, 3). Our corre-
spondent expresses both encouragement and frustration in trying to
communicate how great it is to feel and think as a utopian: "What
I feel is as far beyond human feeling as my thoughts are beyond
human thought. I wish I could show you what I have in mind" (LU,
4). To give us a taste of what a transhuman utopian feels and thinks,
Bostrom offers a plethora of metaphors. These metaphors generally
divide into carrots and sticks. Metaphors of negative reinforcement
are replete with references to pain, punishment, and guilt in charac-
terizing the human condition as akin to poverty, a padlock, prison,

3. Nick Bostrom, "Letter from Utopia," *Studies in Ethics, Law and
Technology* 2, no. 1 (2008): 9. Cited hereafter as LU.

a death trap, a "self-combusting paper hut" (LU, 5), a "loathsome corral" (LU, 6), an assassin, and a "gruesome knot" of suffering (LU, 7). Metaphors of positive reinforcement speak of untold pleasure and knowledge: tears of gratitude, sailing "the high seas of cultures" (LU, 3), finding a secure home, cultivating "nutritious crops of well-being" (LU, 7), "a bubbling celebration of life" (LU, 6). Utopians appeal to us here primarily with passionate hedonism and the magical spice of literary metaphor, rather than rationalistic or scientific argumentation. Yet why are people in Utopia so bipolar? One might hypothesize that transhumans are so equanimous that they would have no need, perhaps even no memory, of these kinds of rhetorical mood swings. Readers have to wonder to what degree this letter is a friendly appeal or manipulative threat, genuinely encouraging or cynically judgmental. Oscillating between lure and condescension will get you "erewhon."

Bostrom consistently rejects arguments from nature or naturalism, yet his letter employs numerous naturalist metaphors and similes. Humanity becoming transhumanity will be "like the full moon that follows a waxing crescent, or like the flower that follows a seed" (LU, 1). We are told that "fun" is "the birth right of every creature" (LU, 6). Also, "All emotions have a natural function" (LU, 7). But how realistic or reliable are these natural cases in the post-natural posthuman condition? Is it necessary to distinguish "given" nature and composed algorithms that can supposedly simulate or enhance any aspect of nature? And what is left of enjoyment of nature as the nonmanufactured or noninstrumentalized environs? Our letter writer basks in countless "astronomical" joys, but strolling among gardens or beaches does not bear mentioning, save perhaps one particularly impenetrable sentence: "I've seen the shoals of colored biography fishes, each one a life story, scintillate under heaving ocean waters" (LU, 3). Here everything in nature is turned up to the sublime just as the metaphorical richness of the sentences are drenched in hyperbole. The sublime itself is collapsed with the pleasure principle: "Every second is so good that it would blow our minds had their amperage not been previously increased" (LU, 9).

A lot of the letter's sentences are evocative of *Blade Runner* Roy Batty–style escaped android statements. As Batty faces death, he tells the bounty hunter Deckard, "I've seen things you people wouldn't believe. Attack ships on fire off the shoulder of Orion. I watched C-Beams glitter in the dark near the Tannhäuser Gate...."[4] Much of the letter carries on in this melancholic lure of "if you could only see what I see." This synthesis of all knowledge and experience in the individual subject is close to what Thierry Bardini describes as *Homo nexus,* a response to infowhelm based on the "nexialist" who can read at the speed of data in A. E. Vogt's novel *The Voyage of the Space Beagle* (1950).[5] Utopian posthumans, however, have pretty much seen everything. Posthumanists not only have time to read everything, they can realize what they read, as if the truth of reading were to become identical with what you read, collapsing reading and being. "I have read all your authors—and much more. I have experienced life in many forms and from many angles: jungle and desert, gutter and palace, heath and suburban creek and city back alley. I have sailed the high seas of cultures, and swum, and dived" (LU, 3). Transhumans are drama queens, sailing the seas of high culture or kitsch, as the case may be. Here art is reduced to enjoyment and self-satisfaction, following what Marcuse called "the affirmative character of culture" and what Leo Bersani describes as the "culture of redemption"[6] that placates the ego above all while eschewing anything disturbing or dissonant. Once the negative, dissenting, critical function of art is dismissed—half the ontology of the artwork—the high seas of kitsch begin to swell.

4. *Blade Runner*'s posthumans would still be existential failures to Bostrom as they have not overcome their own finitude. Ridley Scott, dir., *Blade Runner* (Warner Brothers, 1982).

5. Thierry Bardini, *Junkware* (Minneapolis: University of Minnesota Press, 2011), 145.

6. Herbert Marcuse, *Negations: Essays in Critical Theory,* trans. Jeremy J. Shapiro (London: Mayfly, 2009), 61. Leo Bersani, *The Culture of Redemption* (Cambridge, Mass.: Harvard University Press, 1990).

This is a fantasy of not needing fantasy anymore because there is supposedly no longer a gap between fantasy and reality.

But the "space of the aesthetic," just like the space of existences and reasons, is larger because it includes finitudes, negations, non-systemizations, and precarities. Bostrom's letter promises that only technological enhancement offers this experience of reading as self-realization, but the sentences he offers here are legible on their own because we already have access to aesthetic transformation (not the same as actualized transhumanism) by way of the transfiguring powers of poetics devices such as personification, metaphor, and prolepsis. Forms and genres of the humanities are predicated on such powers inherent in the poetics of language, and these poetic powers do not require special enhancements or technologies. But every document of culture also bears testimony that the materiality of texts, codes of legibility, cultural forms, and transmission practices are themselves uncertain and finite. Texts cannot escape the existentiality of finitude—nothing guarantees that a text will be read or remain legible. Nothing guarantees that a text will be read as its author wants, or that a letter will reach its destination.

Finally, we can raise the question of how irony and deception could play an intractable role in the genre of a letter from a trans-human utopia. How would such a superintelligence address us? How would we know if this address is genuine or a screen behind which lie any number of possible worlds? How would we know if a dystopian calamity or some kind of "desert of the real" stood behind this utopian address? With these questions in mind, the lack of irony and self-critique in this letter is troubling. The letter wants to convince us that posthuman experience is truly wallowing in pleasure, not a simulated experience of pleasure—or rather there is no gap between simulation and reality. But this self-consistency in the form of realized simulation hides a more troubling absence: the self as divided, contradictory, or self-deceptive, possibilities that constitute having a self in the first place. This kind of utopian letter could be written by an artificial intelligence, malicious or not, that is trying to convince humans to commit to the technological path that would end up bringing this artificial intelligence into existence. The letter could be from a dystopian source that is simulating utopia. This letter could just be a machine telling us how great its algorithms are—a scenario that may indeed constitute a utopia to Bostrom. The ultimate irony of this letter is that everything in his utopian imagination is already here and available to us, however brief our lives, but it is unevenly distributed and unevenly shared.

Acknowledgments

I gratefully acknowledge writing this book at Western University situated on the traditional lands of the Anishinaabeg, Haudenosaunee, Lūnaapéewak, and Attawandaron peoples. Research for this book was supported by a grant from the Social Sciences and Humanities Research Council of Canada. I thank my students from two graduate seminars on existential risk taught at Western's Centre for Theory and Criticism in the winter of 2018 and winter of 2020. I learned so much from conversation on this book with my collaborator and cowriter Derek Woods, and from Nicole Shukin, Aaron Brindle, Jeremy Arnott, Julian Evans, Wil Patrick, Kate Stanley, Tilottama Rajan, Allan Pero, Glenn Willmott, Matthew Chrulew, Brett Buchanan, and Stephen Cave. It was a pleasure to have this book under the care of Eric Lundgren and the editorial and production team at University of Minnesota Press. I am deeply thankful for the support, love, and conversations with my parents Stewart and Bette Schuster, my brothers Justin and Jordan Schuster, and my home world of Marina, Reuven, and Raphael Schuster. Marina makes each day sweeter and more utopian.

—JOSHUA SCHUSTER

My portions of this book were written in Lebanon, New Hampshire, on the unceded lands of the Abenaki people, and in Vancouver, British Columbia, on the lands of the Musqueam, Squamish, and

Tsleil-Waututh First Nations. My thanks to the Society of Fellows at Dartmouth College and the Department of English Language and Literatures at the University of British Columbia for their support. I'm also grateful to Yuriko Furuhata, Thomas Patrick Pringle, Yana Stainova, and Evan Morson-Glabik for conversations about the book. Thanks to Dr. Pringle and his students at the University of Chicago for a discussion of existential risk, and to Anne Pasek, Lisa Han, Amy Harris, and the participants of a very sweet pandemic reading group on Media and Environmental Risk. Thanks also to Nicole Shukin, who spoke on the initial panel that shaped this book, and to Eric Lundgren at the University of Minnesota Press. Huge thanks, finally, to my coauthor Joshua Schuster for inspiring this project, inviting me to his existential risk seminar at Western University, and teaching me so much along the way.

—DEREK WOODS

Joshua Schuster is associate professor of English and core faculty member of the Centre for the Study of Theory and Criticism at Western University and author of *The Ecology of Modernism: American Environments and Avant-Garde Poetics.*

Derek Woods is assistant professor of English at the University of British Columbia.

(Continued from page iii)

Forerunners: Ideas First

Davide Panagia
Ten Theses for an Aesthetics of Politics

David Golumbia
The Politics of Bitcoin: Software as Right-Wing Extremism

Sohail Daulatzai
Fifty Years of *The Battle of Algiers*: Past as Prologue

Gary Hall
The Uberfication of the University

Mark Jarzombek
Digital Stockholm Syndrome in the Post-Ontological Age

N. Adriana Knouf
How Noise Matters to Finance

Andrew Culp
Dark Deleuze

Akira Mizuta Lippit
Cinema without Reflection: Jacques Derrida's Echopoiesis and Narcissism Adrift

Sharon Sliwinski
Mandela's Dark Years: A Political Theory of Dreaming

Grant Farred
Martin Heidegger Saved My Life

Ian Bogost
The Geek's Chihuahua: Living with Apple

Shannon Mattern
Deep Mapping the Media City

Steven Shaviro
No Speed Limit: Three Essays on Accelerationism

Jussi Parikka
The Anthrobscene

Reinhold Martin
Mediators: Aesthetics, Politics, and the City

John Hartigan Jr.
Aesop's Anthropology: A Multispecies Approach